Smart Structures and Materials for Sensor Applications

Smart Structures and Materials for Sensor Applications

Guest Editors

Heung Soo Kim
Asif Khan
Jung Woo Sohn

MDPI

Basel • Beijing • Wuhan • Barcelona • Belgrade • Novi Sad • Cluj • Manchester

Guest Editors
Heung Soo Kim
Department of Mechanical, Robotics and Energy Engineering
Dongguk University
Seoul
Republic of Korea

Asif Khan
Department of Mechanical, Robotics and Energy Engineering
Dongguk University
Seoul
Republic of Korea

Jung Woo Sohn
Department of Mechanical Design Engineering
Kumoh National Institute of Technology
Gumi
Republic of Korea

Editorial Office
MDPI AG
Grosspeteranlage 5
4052 Basel, Switzerland

This is a reprint of the Special Issue, published open access by the journal *Sensors* (ISSN 1424-8220), freely accessible at: https://www.mdpi.com/journal/sensors/special_issues/SSMSA.

For citation purposes, cite each article independently as indicated on the article page online and as indicated below:

Lastname, A.A.; Lastname, B.B. Article Title. *Journal Name* **Year**, *Volume Number*, Page Range.

ISBN 978-3-7258-7691-4 (Hbk)
ISBN 978-3-7258-7692-1 (PDF)
https://doi.org/10.3390/books978-3-7258-7692-1

Contents

About the Editors

Heung Soo Kim

Heung Soo Kim received his B.S. and M.S. degrees from the Department of Aerospace Engineering at Inha University, Korea, in 1997 and 1999, respectively. He earned his Ph.D. from Arizona State University in 2003, conducting research supported by NASA Langley Research Center and the US Air Force Research Laboratory on smart composite modeling and computational structural analysis. He is currently a Professor in the Department of Mechanical, Robotics, and Energy Engineering at Dongguk University, Seoul, Republic of Korea. His research interests include the prognostics and health management (PHM) of composite structures, power plants, industrial robots, and mobility batteries. He is also interested in biomimetic actuators, adaptive structures, and structural analysis. He serves as director of the BK21 AIMS Center and the DGU Global Intelligent Robot Center, leading significant research initiatives. He has published 245 SCIE journal papers, achieving an H-index of 46 with over 9,000 citations. His outstanding research contributions have been recognized through numerous Academic Achievement Awards from various academic societies in the PHM field.

Asif Khan

Asif Khan is a Research Scientist at the Mechanics of Composites for Energy and Mobility (MCEM) Laboratory at King Abdullah University of Science and Technology (KAUST), Saudi Arabia. His research focuses on artificial intelligence for structural health monitoring (SHM) of composite structures and rotating machinery, modeling and simulation of defects and rotodynamics, and data augmentation for deep learning-based diagnostics and prognostics. He is also interested in the processing of structural vibrations, guided and Lamb waves, and acoustic emissions for AI-based SHM of non-metallic materials in oil and gas applications, as well as non-destructive testing and evaluation (NDT&E) of non-metallics for energy infrastructure. Within the ENERCOMP consortium on composite asset performance in energy applications, he serves as the Track Leader for Inspection, NDT, and SHM. Asif obtained his Ph.D. in Mechanical Engineering from Dongguk University, Seoul, South Korea, in 2018, and subsequently held academic positions at Dongguk University and the Ghulam Ishaq Khan Institute of Engineering Sciences and Technology. He has authored 22 journal papers, holds two Korean patents, and has presented more than 20 papers at international conferences.

Jung Woo Sohn

Jung Woo Sohn received his B.S. and M.S. degrees from the Department of Mechanical Engineering at Inha University, Korea, in 2001 and 2003, respectively. He received his Ph.D. from Inha University in 2008, conducting research on design and active vibration control of mechanical systems featuring smart materials. He is currently a Professor in the School of Mechanical Engineering at Kumoh National Institute of Technology, Gumi, Republic of Korea. His research interests include AI-based Robot control, including reinforcement learning and imitation learning. He is also interested in the prognostics and health management (PHM) of composite structures.

Preface

Sensors play a vital role across a wide range of industrial and engineering domains, enabling critical functions such as structural health monitoring, process control, condition assessment, and the mitigation of noise and vibration. As modern systems evolve toward higher efficiency, autonomy, and safety, the demand for intelligent sensing technologies has grown substantially. Smart materials—such as piezoelectric ceramics and polymers, shape-memory alloys, electroactive polymers, and magnetostrictive materials—have emerged as key enablers of these technologies due to their inherent ability to sense, actuate, and adapt to changing environmental or structural conditions.

This Special Issue brings together recent advances in the design, modeling, fabrication, and application of sensors based on such smart materials. Theoretical frameworks, numerical simulations, and experimental investigations alike contribute to understanding and optimizing their performance in complex systems. Representative applications include, but are not limited to, piezo-bonded laminated composites for damage detection, morphing wings for adaptive aerodynamics, and reconfigurable engine nozzle fan chevrons for noise reduction.

By integrating multidisciplinary perspectives—from materials science and mechanics to electronics and data analytics—this collection offers a comprehensive view of how smart structures and materials are shaping the next generation of sensor technologies. This Special Issue aims to disseminate new, original, and collective knowledge, gathered from esteemed researchers and practitioners worldwide, on the use of smart structures and materials for sensor applications.

Heung Soo Kim, Asif Khan, and Jung Woo Sohn
Guest Editors

sensors

MDPI

Article

A Deep Learning Framework for Vibration-Based Assessment of Delamination in Smart Composite Laminates

Asif Khan [1], Jae Kyoung Shin [1,2], Woo Cheol Lim [1], Na Yeon Kim [1] and Heung Soo Kim [1,*]

[1] Department of Mechanical, Robotics and Energy Engineering, Dongguk University-Seoul, 30 Pil-dong 1 Gil, Jung-gu, Seoul 04620, Korea; khanuet11@gmail.com (A.K.); jkshin@onepredict.com (J.K.S.); luc2978@dgu.edu (W.C.L.); nayeunk@naver.com (N.Y.K.)

[2] OnePredict Co., Ltd Solution I R&D Division, Seoul 08826, Korea

* Correspondence: heungsoo@dgu.edu; Tel.: +82-2-2260-8577; Fax: +82-2-2263-9379

Received: 26 February 2020; Accepted: 17 April 2020; Published: 20 April 2020

Abstract: Delamination is one of the detrimental defects in laminated composite materials that often arose due to manufacturing defects or in-service loadings (e.g., low/high velocity impacts). Most of the contemporary research efforts are dedicated to high-frequency guided wave and mode shape-based methods for the assessment (i.e., detection, quantification, localization) of delamination. This paper presents a deep learning framework for structural vibration-based assessment of delamination in smart composite laminates. A number of small-sized (4.5% of total area) inner and edge delaminations are simulated using an electromechanically coupled model of the piezo-bonded laminated composite. Healthy and delaminated structures are stimulated with random loads and the corresponding transient responses are transformed into spectrograms using optimal values of window size, overlapping rate, window type, and fast Fourier transform (FFT) resolution. A convolutional neural network (CNN) is designed to automatically extract discriminative features from the vibration-based spectrograms and use those to distinguish the intact and delaminated cases of the smart composite laminate. The proposed architecture of the convolutional neural network showed a training accuracy of 99.9%, validation accuracy of 97.1%, and test accuracy of 94.5% on an unseen data set. The testing confusion chart of the pre-trained convolutional neural network revealed interesting results regarding the severity and detectability for the in-plane and through the thickness scenarios of delamination.

Keywords: delamination; smart composite laminates; structural vibration; spectrograms; deep learning

---

1. Introduction

Composite materials are continuously replacing conventional metallic materials in a variety of industries (e.g., aerospace, automotive) due to their lightweight, high specific strength, high specific stiffness, and design flexibility. Besides, composite materials do not present the typical corrosion problems of conventional metallic materials [1–4]. However, owing to their anisotropic characteristics, laminated composite materials suffer from a variety of complex manufacturing defects and in-service damages such as air trapped between plies, non-uniform distribution of epoxy, inadequate consolidation pressure, delamination, matrix cracking, and fiber fracture, among others [5–8]. The presence of defects in laminated composites is more critical in terms of effects and more challenging in terms of assessment (i.e., detection, quantification, and localization) than those in the metallic materials. The presence of delamination or separation between plies is one of the most detrimental defects in the laminated composite as it can cause up to 60% loss in the strength without any perceivable effects at the surface or noticeable change in the stiffness [9,10]. Once initiated, delamination could propagate into a wide damage zone and thus compromising the structural integrity of the component or structure [11,12]. To ensure safe and reliable operation of laminated composite materials in real engineering structures, it is imperative to timely identify, quantify, and localize the presence of delamination in these materials.

The anisotropic material properties and complex microstructure make the damage assessment of composite materials more challenging than metallic materials. Consequently, the damage assessment of laminated composite is an active research area where continuous efforts are made to find cost-effective techniques for the detection, quantification, and localization of delamination in these materials [13–16]. Various non-destructive testing (NDT) procedures are being used for the assessment of delamination such as acoustic emission, thermography, ultrasonic, and X-ray [17–21]. However, most of the NDT techniques are often costly, labor-intensive, unable to handle large size test objects, and depend heavily on the expertise of the operator. Hence, continuous research efforts are underway to come up with new techniques for the damage assessment of laminated composites. Zhao et al. [22] proposed a local wavenumber method for the localization and characterization of mode 1 delamination in carbon fiber/epoxy composite laminates. The wave propagation velocity was found to transform at the delamination and an increase in the central frequency was found to cause larger attenuation and dispersion in the guided waves. Grassia et al. [23] developed a strain-based method for structural health monitoring of composite structures by making a fingerprint model of the reference healthy structure via a neural network. The developed method was able to detect damages from the early stage of micro-cracks up to delamination in composite structures. Yang et al. [24] proposed that the variations of modal frequency under mass loading could be used to quantify the effects of delamination in laminated composites. Mei et al. [25] studied high-frequency local vibration for detecting and quantifying the size, shape, and depth of delamination in composite plates. Sikdar et al. [26] investigated the effects of debonding and variable ambient temperature on the propagation of Lamb wave in composite materials. The presence of delamination was reflected by a reduction in the amplitudes and velocity of A_0 mode of the Lamb wave. The increase in temperature further decreased the propagation velocity and amplitudes of A_0 mode. Temperature correction factors were proposed to minimize the effect variable temperature conditions on structural health monitoring strategies for composite materials. Kundu et al. [27] proposed a generic machine learning framework for the detection and localization of damage in composite materials using spatially and temporally correlated features of an acoustic emission signal. Khan et al. [28] presented a machine learning paradigm for the detection, quantification, and prediction of delamination in smart composite laminates. Pre-trained predictive models were found to predict physically reasonable labels for new unseen cases of delamination. Chen et al. [29] used a laser Doppler vibrometer and curvatures of the operating deflection shapes (CODS) for the detection and location of delamination in laminated composites. It was found that third and fifth pure bending modes show more sensitivity to the presence of local delaminations. Yelve et al. [30] proposed a Lamb wave-based nonlinear method for the assessment of delamination in laminated composites. A spectral damage index was extracted from the fundamental and higher harmonics for assessing the size of delamination, while spectral and temporal data were used for the localization of delamination. Feng et al. [31] studied the propagation and dispersion of Lamb waves in delaminated composite laminates. It was found that the delamination length could be quantified from the phase difference of two waves travelling in the upper and lower sub-laminates. Dafydd et al. [32] investigated the potential of ultrasonic guided waves for the severity assessment of impact damage in carbon-fiber-reinforced polymers. Among the two fundamental modes of A_0 and S_0, A_0 mode showed more sensitivity to the damage severity than S_0. Li et al. [33] developed a numerical model for the propagation and dispersion of ultrasonic guided waves in damaged composite using high-fidelity local interaction simulation. The time-frequency features obtained via the matching pursuit decomposition algorithm showed that a signal from symmetric excitation was more sensitive to damage location than the signal from the anti-symmetric excitation. Huang et al. [34] studied principal component analysis (PCA) for the detection and quantification of delamination in smart composite laminates. It was found that distance between the healthy and delaminated data clusters could be used to quantify the severity of delamination.

In general, the vibratory responses in the fundamental modes (e.g., structural vibration) carry information about the presence of damage (i.e., global behavior) in the structure, whereas information on the location of damage is captured by high-frequency Lamb waves and acoustic emission [35]. In the published literature on damages in laminated composites, the presence of delamination had been assessed from the structural vibration and modal parameters [36], whereas guided waves and acoustic emission had been used for the detection and localization of delamination [37,38]. However, some practical limitations of the methods that employ high-frequency waves for nondestructive evaluation are their need of too many sensors, optimized location of the receivers, the damage to be between the actuator(s) and sensor(s), data acquisition at a higher rate, and complex signal processing [39–41]. On the contrary, structural vibration is more readily available for the assessment of damage due to its easy measurement via smart materials [42]. In addition, the delamination is not necessary to be in the path between the sensor and actuator, unlike the guided wave-based methods. Hence, in this work, an effort has been made to use low-frequency structural vibration for the local and global assessment of inner and edge delaminations in laminated composites.

From the contemporary research on delamination in laminated composites, it can be found that most of the research efforts are dedicated to the use of higher-frequency guided waves (e.g., Lamb waves), acoustic emission/acoustic ultrasonic, and mode shape curvatures for the detection, quantification, and localization of delamination [43–49]. This paper proposes a deep learning framework for the assessment of delamination in piezo-bonded laminated composites using low-frequency structural vibration responses. An electromechanically coupled mathematical model of the laminated composite with piezoelectric sensors and actuators is developed and solved in the time domain. Several cases of inner and edge delaminations are simulated in a smart cantilever plate. The transient responses of the healthy and delaminated smart composite laminates are transformed into spectrograms, which are processed with the deep convolutional neural network (CNN). Contrary to conventional hand-crafted discriminative features, the CNN automatically extracts discriminative features from the spectrogram and uses those features to classify the healthy and delaminated smart composite laminates. The obtained results showed that the proposed approach could be employed for the assessment of inner and edge delaminations in smart composite laminates while using low-frequency structural vibration only.

2. Problem Formulation

The accurate description of the structural deformation of a laminated composite with surface bonded or embedded piezoelectric sensors and actuators requires a coupled electromechanical formulation. The kinematics of the smart structure with single and multiple discrete delaminations is developed based on improved layerwise theory [50], whereas higher-order electric potential function [51,52] is used to model the potential variation through the piezoelectric patches. As per the improved layerwise theory, first-order shear deformation theory is superimposed with layerwise functions to describe the displacements of a point with coordinates (x, y, z) as follows [34]:

$$\begin{aligned} U_i^k(x,y,z,t) = & \; u_i(x,y,t) + A_i^k\psi_1 + B_i^k\psi_2 + C_i^k w_{,x} + D_i^k w_{,y} \\ & + E_i^k\{\overline{w}^j_{,x}\} + F_i^k\{\overline{w}^j_{,y}\} + \sum_{j=1}^{N-1} \overline{u}_i^j H(z-z_j) \\ U_3^k(x,y,z,t) = & \; w(x,y,t) + \sum_{j=1}^{N-1} \overline{w}^j H(z-z_j) \end{aligned} , \quad (1)$$

where $U_i^k (i = 1,2)$ and U_3^k refer to the in-plane displacements and transverse deflection, respectively. The terms $u_i(i = 1,2)$ and w respectively denote the in-plane and transverse displacements at the mid-plane of the laminate. The symbol $\psi_i(i = 1,2)$ describes the rotation of normal to the mid-plane, and thus accounts for the shear stress variation through the thickness of the laminate. The terms of $\overline{u}_i^j$

and $\overline{w}^j$ account for the possible slipping and jump in the displacements at the debonded interfaces between adjacent plies, H denotes Heaviside function, and z_j refers to the delaminated interface. The terms $A_i^k, B_i^k, C_i^k, D_i^k, E_i^k$, and F_i^k are layerwise coefficients which are calculated in terms of the material and geometric properties of the laminates [50]. In the displacement field of Equation (1), perfectly bonded and delaminated interfaces are modeled by setting the terms $\overline{u}_i^j$ and $\overline{w}^j$ to zero and nonzero, respectively.

The electric potential field of the piezoelectric sensors and actuators is assumed to have a cubic distribution along the thickness direction. The mathematical description of the potential field for the p-th piezoelectric layer is given by Equation (2) [51]:

$$\begin{aligned}\phi^p(x,y,z,t) &= \phi_0^p(x,y,t) - \left(z - z_0^p\right)E_z^p(x,y,t) + 4\left(\frac{z-z_0^p}{h^p}\right) \\ &\times\left[\left(z - z_0^p\right)\left(\frac{\overline{\phi}^p(x,y,t)}{h^p} + E_z^p(x,y,t)\right) - \phi_0^p(x,y,t)\right]\end{aligned}, \tag{2}$$

where ϕ_0^p and E_z^p refer to the electric potential and electric field at the mid-plane of the p-th piezoelectric layers, respectively. The symbol $\overline{\phi}^p$ shows the potential difference between the top and bottom electrodes, z_0^p and h^p refer to the mid-plane position and thickness of the piezoelectric patch, respectively. The 2nd and 3rd terms of Equation (2) respectively denote the linear and nonlinear potential variation through the thickness of the piezoelectric layer.

In the current work, linear constitutive relations (i.e., constant material coefficients) are considered for the converse and direct piezoelectric effects. The constitutive relations of piezoelectric actuator and sensor are shown by Equation (3) [51]:

$$\begin{aligned}\sigma_{ij} &= c_{ijkl}\varepsilon_{kl} - e_{ijk}E_k \\ D_i &= e_{ijk}\varepsilon_{jk} + b_{ij}E_j\end{aligned}, \tag{3}$$

where σ_{ij} and D_i denote the stress tensor and electric displacement vector, respectively. The quantity ε_{ij} and E_i respectively denote the strain tensor and the electric field vector. The terms c_{ijkl} and e_{ijk} refer to the elastic and piezoelectric constants, respectively. The term b_{ij} denotes the dielectric permittivity of the piezoelectric material. For the linear piezoelectricity of Equation (3), the electric field vector E_i is obtained from the scalar potential field of Equation (2) as follows:

$$E_i = -\phi_{,i}\ (i = 1,2,3), \tag{4}$$

where the subscript denotes partial derivative with respect to i with $i = 1, 2, 3$.

A finite element method is adopted to combine the displacement and potential fields for a 4-noded plate element. The primary in-plane unknowns $(u_1, u_2, \psi_1, \psi_2, \overline{u}_1^j, \overline{u}_2^j)$ of the displacement field and electrical unknowns (ϕ_0^p, E_z^p) of the potential field are described in terms of linear Lagrange interpolation functions, whereas Hermite cubic interpolation functions are used for out-of-plane unknowns of the displacement field, as shown by Equation (5) [51]:

$$\begin{aligned}&\left(u_1,u_2,\psi_1,\psi_2,\overline{u}_1^j,\overline{u}_2^j\right) = \sum_{m=1}^{4} N_m\left[(u_1)_m,(u_2)_m,(\psi_1)_m,(\psi_2)_m,\left(\overline{u}_1^j\right)_m,\left(\overline{u}_2^j\right)_m\right] \\ &\left(\phi_0^p, E_z^p\right) = \sum_{m=1}^{4}\left[N_m\left(\phi_0^p\right)_m, N_m\left(E_z^p\right)_m\right] \\ &w = \sum_{m=1}^{n}\left[H_m(w)_m + H_{xm}(w_{,x}) + H_{xm}\left(w_{,y}\right)_m\right] \\ &\overline{w}^j = \sum_{m=1}^{n}\left[H_m(\overline{w})_m + H_{xm}(\overline{w}_{,x}) + H_{xm}\left(\overline{w}_{,y}\right)_m\right]\end{aligned}, \tag{5}$$

where N_m refers to Lagrange interpolation function and H_m, H_{xm}, H_{ym} are Hermite interpolation functions.

From Equation (5), the displacement unknowns can be described in terms of nodal unknowns by using the matrix notation as follows:

$$\begin{array}{c}\{u_u^e\} = [N_u]\{d_u\} \\ \{u_\phi^e\} = [N_\phi]\{d_\phi\}\end{array}, \tag{6}$$

where

$$\begin{array}{l}\{u_u^e\} = \left[u_1, u_2, w, \psi_1, \psi_2, \overline{u}_1^j, \overline{u}_2^j, \overline{w}^j\right]^T \\ \{d_u\} = \left[\cdots, u_{1i}, u_{2i}, w_i, w_{,xi}, w_{,yi}, \psi_{1i}, \psi_{2i}, \overline{u}_{1i}^j, \overline{u}_{2i}^j, \overline{w}_{,i}^j \overline{w}_{,xi}^j, \overline{w}_{,yi}^j, \cdots\right] \\ \{u_\phi^e\} = \left[\phi_0^p, E_z^p\right]^T \\ \{d_\phi\} = \left[\cdots, \phi_{0i}^p, E_{zi}^p, \cdots\right]^T\end{array}. \tag{7}$$

In terms of the displacement field (Equation (1)) and electric potential field (Equation (2)), the elemental displacement field $u(x,y,z,t)$, elemental strain field $\varepsilon(x,y,z,t)$, and electric potential field $E^p(x,y,z,t)$ can be described as follows:

$$\begin{array}{l}u(x,y,z,t) = L_u u_u^e(x,y,t) \\ \varepsilon(x,y,z,t) = L_\varepsilon u_u^e(x,y,t) \\ \phi^p(x,y,z,t) = V_b\left(z - z_0^p, h^p, \overline{\phi}^p\right) + L_\phi^p u_\phi^p(x,y,z,t) \\ E^p(x,y,z,t) = -F_b\left(z - z_0^p, h^p, \overline{\phi}^p\right) - L_E^p u_\phi^j(x,y,z,t)\end{array}, \tag{8}$$

where the operators L_u. $L_\varepsilon, L_\phi^p, L_E^p$ and the expressions for V_b and F_b are shown in Appendix A. The equation of motion is obtained by using the variational principle as shown by Equation (9):

$$\begin{array}{l}\delta\pi_u = \int_0^t \left[\int_V \left(\rho\ddot{u}_i\delta u_i + \sigma_{ij}\delta\varepsilon_{ij} + \xi\dot{u}_i\delta u_i\right)dV + \int_S t_i\delta u_i dS\right]dt = 0 \\ \delta\pi_\phi = \int_0^t \left[\int_V D_i\delta\phi_{,i}dV + \int_S q_e\delta\phi dS\right]dt = 0\end{array}, \tag{9}$$

where $\delta\pi_u$ and $\delta\pi_\phi$ refer to the energy functional of the mechanical and electrical fields, respectively. The terms ρ denotes mass density, ξ is damping ratio, t_i is traction vector, and q_e represents charge density.

The electromechanically coupled equation of motion is obtained by substituting the stress, strain, and electric displacement components in Equation (9), and is shown in matrix form by Equation (10) [51]:

$$\begin{array}{r}M\ddot{d}_u + C\dot{d}_u + K_{uu}d_u + K_{u\phi}d_\phi = F_u \\ K_{\phi u}d_u + K_{\phi\phi}d_\phi = F_\phi\end{array}, \tag{10}$$

where M and C respectively denote structural mass and damping matrices. The terms K_{uu} and $K_{\phi\phi}$ refer to the stiffness matrices of the mechanical and electrical fields, respectively. The symbols of $K_{u\phi}$ and $K_{\phi u}$ show the electromechanical coupling matrices, the vectors d_u and d_ϕ refer to the displacement unknowns and electrical unknowns of piezoelectric patches. The terms F_u and F_ϕ denote the applied mechanical and electrical forces, respectively. The actuation and sensing mechanism of the piezoelectric actuator and sensors is accounted for by the piezoelectric–mechanical coupling matrices of $K_{u\phi}$(converse piezoelectric effect) and $K_{\phi u}$(direct piezoelectric effect). The coupling matrices allow the piezoelectric actuator to produce mechanical actuation forces under input voltages and the piezoelectric sensors to generate electrical signals under mechanical deformation.

For numerical solution, the mathematical model of Equation (10) is modified to Equation (11) by applying matrix condensation:

$$M\ddot{d}_u + C\dot{d}_u + \left(K_{uu} - K_{u\phi}K_{\phi\phi}^{-1}K_{\phi u}\right)d_u = F_u - K_{u\phi}K_{\phi\phi}^{-1}F_\phi. \quad (11)$$

The developed numerical model was implemented in MATLAB. Newmark's time integration algorithm [34] was employed to solve the electromechanically coupled model of Equation (11) in the time domain. In the algorithm, Taylor's expansions with terms up to the second derivative were used to approximate the displacement variable, its first and second derivatives. The governing equation was reduced to a set of algebraic equations that were solved through an iterative process involving the force input at each time step and the initial conditions.

3. Numerical Example

The mathematical formulation of Section 2 is numerically implemented on a smart laminated composite plate with various cases of small sized (4.5% of the total area) inner and edge delaminations. The smart plate is made of 16-layers which are stacked together in a symmetric cross-ply configuration ($[0/90]_{4s}$). One piezoelectric actuator and three piezoelectric sensors are attached to the surface of the smart plate as shown in the schematic of Figure 1.

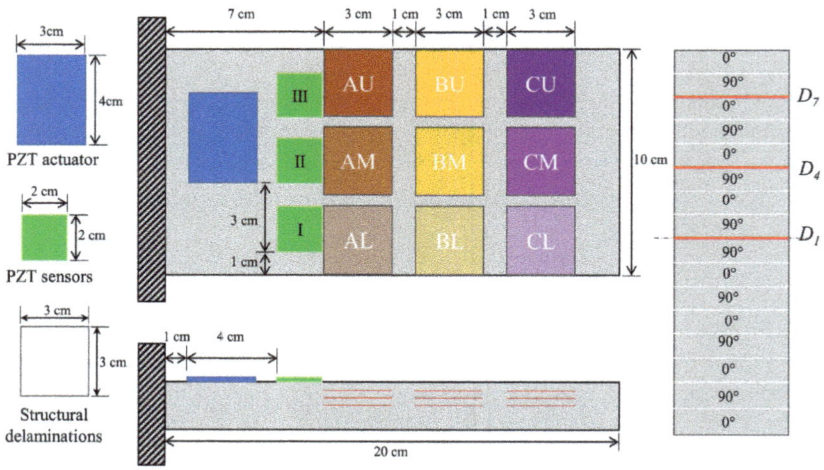

Figure 1. Schematic of smart composite laminate with various inner and edge delaminations (**left**) Top and front views (**right**) exaggerated view of the thickness direction.

Herein, the piezoelectric actuator is attached at the off-center position along the width direction so that mix-vibrating modes (i.e., bending and twisting) of the smart plate could be stimulated. The piezoelectric actuator and sensors are made of the same materials. Tables 1 and 2 respectively show the material properties of a single lamina of the laminated composite and piezoelectric material.

Table 1. Material properties of a single lamina of the laminated plate.

E_1	E_2, E_3	G_{12}, G_{13}	G_{23}	ρ	ν_{12}, ν_{13}	ν_{23}
372 GPa	4.12 GPa	3.99 GPa	3.6 GPa	1788.5 kg/m^3	0.275	0.42

Table 2. Material properties of the piezoelectric sensors and actuator.

E	ν	ρ	d_{31}, d_{32}	d_{24}, d_{15}	d_{36}
69 GPa	0.31	7700 kg/m^3	179×10^{-12} C/N	-741×10^{-12} C/N	0

In Figure 1, the symbols *AM*, *BM*, and *CM* denote the inner delaminations, whereas *AL*, *AU*, *BL*, *BU*, *CL*, and *CU* refer to the cases of edge delaminations. The inner and edge delaminations could occur at three different interfaces of D_1 (mid-plane), D_4, and D_7 through-the-thickness of the smart plate. The in-plane delaminations combined with through-the-thickness interface are named by adding the respective interface number with the name of the in-plane location. For example, *AL1* means the presence of delamination at the *AL* in-plane location and at the D_1 interface. The smart plate without any defect is referred to as the '*Healthy* (*H*)' case. The nine in-plane delaminations combined with three options of through-the-thickness interfaces add up to 27 cases of delaminations in the smart plate.

In the FEM model of the smart plate, the host laminate and the piezoelectric patches were discretized with four-node plate elements. The healthy and delaminated smart plates were excited into random low-frequency vibration through a set of 1000 random harmonic voltages applied through the piezoelectric actuator and the corresponding responses were measured through the piezoelectric sensors. Figure 2a shows the response spectrum of *AL1* to the 5 random loadings as measured by sensor I, whereas Figure 2b shows the frequency response spectrum of *H*, *AL1*, *AM1*, and *AU1* to one of the random inputs as measured by sensor I.

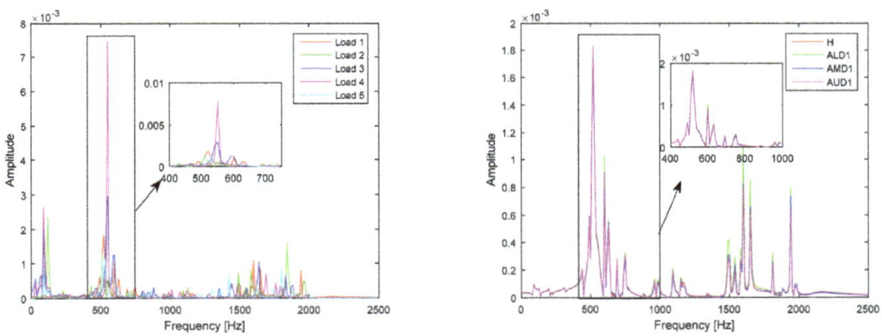

(**a**) (**b**)

Figure 2. Frequency spectrum of the transient response for: (**a**) a single case (*AL1*) to 5 random loadings; (**b**) Healthy and different delaminated cases to a single random load.

The frequency spectrum of the transient responses in Figure 2a shows that random loadings stimulated multiple modes of vibration and caused more variability in the transient response of a single case (*AL1*). On the other hand, the variations caused by different damage cases (Figure 2b) is very subtle for different cases of in-plane delaminations and would further decrease for different interfaces of delaminations. The comparison of Figure 2 shows that conventional time and frequency domain discriminative features (e.g., mean, kurtosis, variance, skewness, etc.) could not be employed to discriminate the healthy plate from the delaminated plate.

In general, the variations of modal parameters (i.e., natural frequency, damping, vibration mode shapes) are commonly used for the detection of delamination in the interlaminar regions of composite materials. However, model parameters have been mostly used for the detection of delamination and limited research effort can be found on the use of variation in modal parameters for the detection, quantification, and localization of delamination. Besides, monitoring variations in the modal damping and measurement of vibration mode shapes for delamination detection are challenging tasks, especially in practical experiments. In addition, the variations of the natural frequencies usually reflect the effect of large size delamination, and smaller delamination may not show any noticeable variation in the natural frequencies. Although, from Figure 2b, certain cases of delamination may show some difference of amplitude in comparison with the healthy case, the same trend could not be observed for all cases of delamination due to difference in the severity levels of different delamination. The less severe cases of delamination (e.g., AMD7, BMD7, CMD7, etc.) would not show any appreciable difference for the healthy and delaminated smart composite laminates. In addition, the variations in Figure 2b are smaller than in Figure 2a and the changes in natural frequency could not be used for the assessment of inner and edge delaminations in the laminated plate.

4. Proposed Methodology

The frequency spectrums in Figure 2 show that the structural vibration responses from the healthy and delaminated smart composite plates are more influenced by operating conditions than the delaminations in the plate. The randomness of loadings hinders the use of statistical time domain and frequency domain features for distinguishing the healthy case from the delaminated cases as well as a distinction between different cases of delaminations (e.g., *AL1*, *AL2*, *AL3*, *AM1*, etc.). Deep learning offers a natural solution to the current problem due to its capability to automatically extract discriminative features from the input images [53]. Figure 3 shows a schematic of the deep learning-based methodology for the current problem.

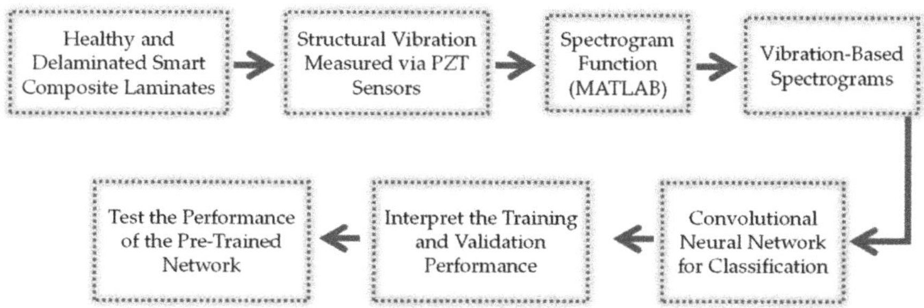

Figure 3. Schematic of the deep learning-based methodology for structural vibration-based delamination assessment of smart composite laminates.

Herein, the response from the three piezoelectric sensors is transformed into time-frequency spectral images using the *spectrogram* function in Matlab. The spectrogram function allows for controlling the time and frequency resolution of the spectrograms by adjusting the variables of *window size*, type of window function (*Hamming*, *Hann*, and *Black* window function), fast Fourier transform *(FFT) resolution*, and the *overlapping rate*. In general, the FFT resolution was observed to affect the spectral resolution of the vibration spectrograms and the computational time of the short time Fourier transform. In the current work, the FFT resolutions of 10, 10^2, 10^4, and 10^8 were considered for initial assessment and the spectrogram with FFT resolution of 10^4 and 10^8 showed good spectral resolution. Although no significant difference was observed in the resolution of spectrograms with 10^4 and 10^8 FFT resolutions, the computational time for 10^8 FFT resolution was about 3 times longer than for 10^4 FFT resolution. Hence, an FFT resolution of 10^4 was selected. The Hamming window function was selected due to its better performance in the high-frequency band compared with Hann and Blackman window functions. To prevent Picket fence error and avoid loss of information at the beginning and end of the signals [54], a 50% overlapping range was selected for the window function. In order to check the size of the window, which is the remaining resolution parameter, the spectral images were generated by specifying the size of the window as 10, 20, and 50. It was observed that smaller window size resulted in high time resolution and low frequency resolution, and vice versa. Hence, the size of the window was set to 20 to obtain a spectral image with an appropriate balance of time and frequency resolution. For the convolutional neural network (CNN), the spectrograms of the three sensors for a single loading were staked and supplied as nine-channel input (3 RGB channels for each sensor) as shown in Figure 4.

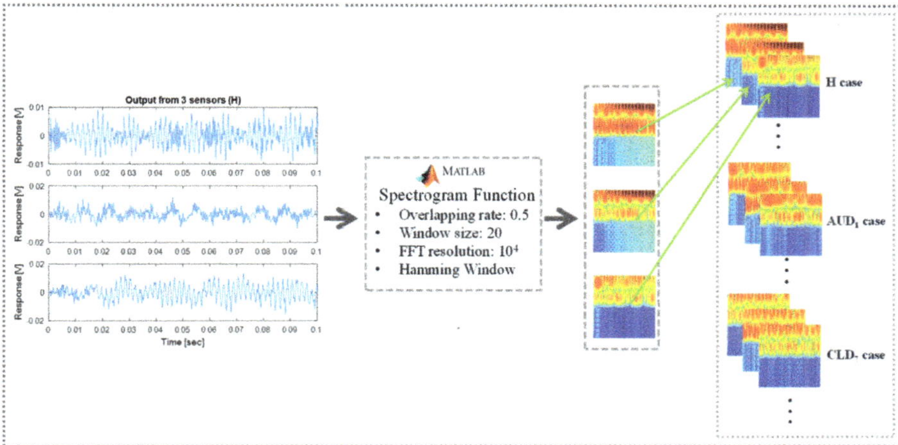

Figure 4. Preparation of vibration-based spectrograms for deep learning using the spectrogram function of Matlab.

The original size of each spectrogram was 768 × 512 × 3 (width × height × number of channels) with the pixel values between 0 and 255. To reduce the computation cost of the deep learning algorithm and remove the divergence from the pixel values, each one of the three spectrograms was reduced to 256 × 256 × 3 (width × height × number of channels) and normalized to a pixel value between 0 and 1 as shown by Figure 5.

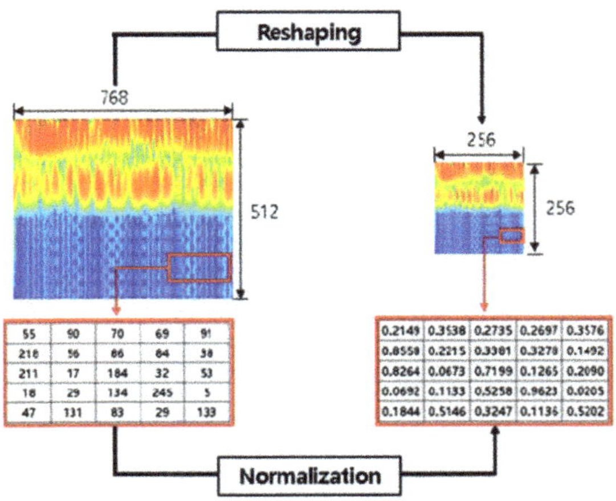

Figure 5. Size conversion and normalization of spectral images.

The goal of normalization was to change the pixel values of the spectrograms to a common scale for the healthy and delaminated structures without distorting the differences in the range of values. In addition, normalization reduced the influence of larger variance in the pixel values on the classification results of the deep learning algorithm. Details of the CNN-based classifier, interpretation of the training and validation process, and testing of the pre-trained network are discussed in the next section.

5. Results and Discussion

The data set of vibration-based spectrograms consisted of one healthy case (*H*) and 27 cases of inner and edge delaminations in the smart laminated composite plate. Based on the applied random harmonic loads, each one of the 28 cases is comprised of 1000 instances. From the discussion in Section 4, each instance consists of 3 RGB spectrograms for one random load. The 28,000 spectrograms were split into 22,400 (80%) spectrograms for training, 2800 (10%) for validation of the network during training, and 2800 (10%) were used for evaluating the testing performance of the pre-trained network. A block diagram of the convolutional neural network used for the classification of vibration-based spectrograms is shown in Figure 6.

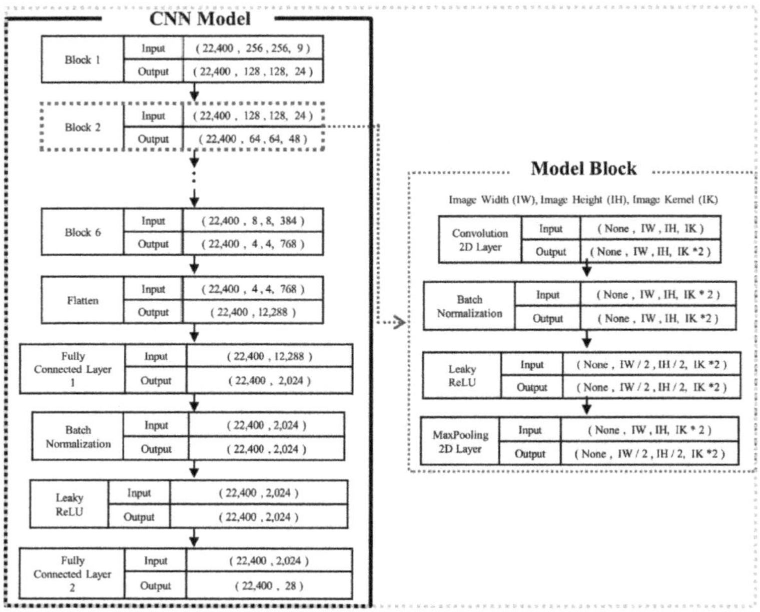

Figure 6. Architecture of the convolutional neural network for the delamination assessment in smart composite laminates.

Herein, the input to the network consists of 22,400 training spectrograms where the size of each spectrogram is equal to $256 \times 256 \times 9$ (width × height × number of channels). In the network, six model blocks were used to automatically extract discriminative features from the spectrograms. Each model block was comprised of a 2D convolution layer with a filter/kernel of size 3×3, batch normalization, Leaky ReLU [55,56] as a nonlinear activation function, and a 2D max-pooling with a filter of size 2×2. Batch normalization and 2D max-pooling layers were used in each model block to prevent the network from overfitting of the training data. The kernel size of 3×3 was kept constant for all the 2D convolutional layers of the six model blocks. Whereas, the number of kernels in the six model blocks was respectively chosen to be 24, 24×2, 24×4, 24×8, 24×16, and 24×32. The three-dimensional output (4, 4, 768) from the last model block is converted into one-dimensional data (12,288) using a Flatten layer. After that, the data is passed through a fully connected hidden layer with 2024 neurons, batch normalization, the activation function of Leaky ReLU. The features are finally passed to a fully connected output layer with 28 neurons. The mathematical details and functionality of each layer of the CNN can be referred to in the references [57,58]. In the last fully connected layer of the network, a softmax function with a built-in function of TensorFlow (i.e., softmax_cross_entropy_with_logits_v2) was used to classify the 28 scenarios of the healthy and delaminated smart composite laminates.

The learnable parameters of weights and bias were randomly initialized and trained from scratch. The hyperparameters of learning rate, number of hidden layers, and size and number of convolutional kernels were decided through an empirical experiment for the given dataset of healthy and delaminated smart composite laminates. The proposed CNN was trained up to 200 epochs with a learning rate of 0.0001 and Adam optimizer was used for fast calculation. CNN architecture was implemented using the open-source platform of TensorFlow on a GeForce RTX 2080 Ti GPU from NVIDIA.

The proposed architecture of the CNN was trained through a 10-fold cross-validation technique. The network showed an overall all training accuracy of 99.9% and validation accuracy of 97.1%. The high validation accuracy showed that the network had not overfitted the training data.

To evaluate the performance of the trained network, the pre-trained network was tested with 2800 spectrograms that were not used during the training and validation process. The pre-trained network showed a predictive test accuracy of 94.5% and Figure 7 depicts per-class predictive performance in the form of a confusion chart.

	H	AU1	AU4	AU7	AM1	AM4	AM7	AL1	AL4	AL7	BU1	BU4	BU7	BM1	BM4	BM7	BL1	BL4	BL7	CU1	CU4	CU7	CM1	CM4	CM7	CL1	CL4	CL7
H	100	0	0	0	0	0	0	0	0	0	0	0	0	0	0	0	0	0	0	0	0	0	0	0	0	0	0	0
AU1	0	100	0	0	0	0	0	0	0	0	0	0	0	0	0	0	0	0	0	0	0	0	0	0	0	0	0	0
AU4	0	0	100	0	0	0	0	0	0	0	0	0	0	0	0	0	0	0	0	0	0	0	0	0	0	0	0	0
AU7	0	0	0	100	0	0	0	0	0	0	0	0	0	0	0	0	0	0	0	0	0	0	0	0	0	0	0	0
AM1	0	0	0	0	100	0	0	0	0	0	0	0	0	0	0	0	0	0	0	0	0	0	0	0	0	0	0	0
AM4	0	0	0	0	0	99	0	0	0	0	0	0	0	0	1	0	0	0	0	0	0	0	0	0	0	0	0	0
AM7	0	0	0	0	0	0	92	0	0	0	0	0	1	0	0	7	0	0	0	0	0	0	0	0	0	0	0	0
AL1	0	0	0	0	0	0	0	100	0	0	0	0	0	0	0	0	0	0	0	0	0	0	0	0	0	0	0	0
AL4	0	0	0	0	0	0	0	0	100	0	0	0	0	0	0	0	0	0	0	0	0	0	0	0	0	0	0	0
AL7	0	0	0	0	0	0	0	0	0	100	0	0	0	0	0	0	0	0	0	0	0	0	0	0	0	0	0	0
BU1	0	0	0	0	0	0	0	0	0	0	100	0	0	0	0	0	0	0	0	0	0	0	0	0	0	0	0	0
BU4	0	0	0	0	0	0	0	0	0	0	0	100	0	0	0	0	0	0	0	0	0	0	0	0	0	0	0	0
BU7	0	0	0	0	0	0	1	0	0	0	0	0	93	0	0	6	0	0	0	0	0	0	0	0	0	0	0	0
BM1	0	0	0	0	0	0	0	0	0	0	0	0	0	95	0	0	0	0	0	0	0	0	5	0	0	0	0	0
BM4	0	0	0	0	0	1	0	0	0	0	0	0	0	0	91	0	0	0	0	0	0	1	0	6	0	0	0	1
BM7	0	0	0	0	0	0	8	0	0	0	0	0	5	0	0	87	0	0	0	0	0	0	0	0	0	0	0	0
BL1	0	0	0	0	0	0	0	0	0	0	0	0	0	0	0	0	100	0	0	0	0	0	0	0	0	0	0	0
BL4	0	0	0	0	0	0	0	0	0	0	0	0	0	0	0	0	0	100	0	0	0	0	0	0	0	0	0	0
BL7	0	0	0	0	0	0	0	0	0	0	0	0	0	0	1	0	0	0	94	0	0	0	0	0	2	0	2	1
CU1	0	0	0	0	0	0	0	0	0	0	0	0	0	2	0	0	0	0	0	98	0	0	0	0	0	0	0	0
CU4	0	0	0	0	0	0	0	0	0	0	0	0	1	0	1	0	0	0	0	0	97	1	0	0	0	0	0	0
CU7	0	0	0	0	0	0	0	0	0	0	0	0	0	0	1	1	0	0	1	0	0	84	0	3	6	0	0	4
CM1	1	0	0	0	0	0	0	0	0	0	0	0	0	0	0	0	0	0	0	0	0	0	99	0	0	0	0	0
CM4	0	0	0	0	0	0	0	0	0	0	0	0	0	0	1	1	0	0	1	0	0	2	0	84	6	0	0	5
CM7	0	0	0	0	0	0	0	0	0	0	0	0	0	0	1	1	0	0	3	0	0	6	0	10	68	0	0	11
CL1	0	0	0	0	0	0	0	0	0	0	0	0	0	0	0	0	0	0	0	0	0	0	0	0	0	100	0	0
CL4	0	0	0	0	0	0	0	0	0	0	0	0	0	0	0	0	0	0	1	0	0	0	0	0	0	0	99	0
CL7	0	0	0	0	0	0	0	0	0	0	0	0	0	0	1	0	0	0	12	0	0	2	0	5	12	0	2	66

Figure 7. Confusion matrix of the pre-trained convolutional neural network (CNN) on unseen test data.

Herein, the labels in the first column denote the ground truth or actual classes, the topmost row shows the labels predicted by the pre-trained convolutional neural network, the numbers on the main diagonal represent the percentage of correctly classified instances, and the off-diagonal shows the percentage of incorrectly classified instances. It is worth noting that delamination at the mid-plane interface shows more noticeable effects in terms of stiffness change and dynamic response characteristics than delamination of the same size near the free surfaces (away from mid-plane) [50,59]. In addition, due to the high strain near the fixed end of the cantilever plate and proximity of the piezoelectric sensors, delamination near the fixed end would more pronouncedly affect the dynamic response than delamination of the same size near the free end of the smart plate. The test confusion matrix of Figure 7 reveals the following results:

1. The classifier has distinguished the healthy case from the delaminated cases with 100% accuracy.
2. The more severe cases of delaminations (the one that occurs at the mid-plane interface) have been identified with 95%~100% accuracy.
3. The pre-trained network can distinguish the inner delamination (*AM*, *BM*, and *CM*) from the edge delaminations (*AL*, *AU*, *BL*, *BU*, *CL*, and *CU*) with 90%~100% accuracy.
4. The major loss of accuracy is due confusion between the least severe cases of delaminations (i.e., inner and edge delaminations that occur near the free surface and whose position is furthest from the sensors). More specifically, the smallest accuracy has been observed for *CL7* (66%) and

CM7 (68%). In case of *CL7*, the misclassification results are 2% as *CL4* (same in-plane location, different interface of delamination), 12% as *CM7* (same interface, different in-plane location), 5% as *CM4* (different in-plane location, different interface), 2% as *CU7* (same interface, different in-plane location), 12% as *BL7* (same interface, different in-plane location), and 1% as *BM4* (different in-plane location, different interface). Herein, major misclassification is due to confusion between delaminations at the less severe interface along the thickness i.e., D_7. Same is the case for misclassifications of *CM7*.

The pre-trained CNN model classified the healthy and delaminated smart composite laminates with a classification accuracy of 94.5%. To get an idea of the classification performance regarding the in-plane location (*A*, *B*, *C*) and through-the-thickness interface (D_1, D_4, D_7) of delamination, the classification accuracy is averaged according to the in-plane location and through-the-thickness interface as shown by Figure 8.

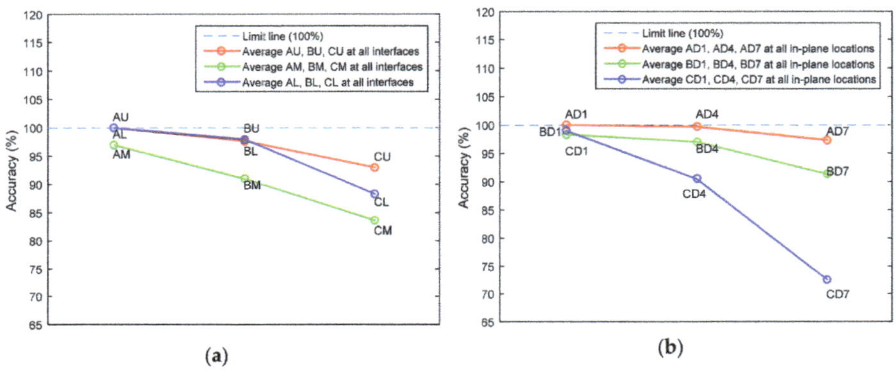

Figure 8. Average predictive performance of the pre-trained CNN with respect to: (**a**) in-plane location of delamination; (**b**) through-the-thickness interface of delamination.

For the in-plane location of delamination (Figure 8a), it is observed that the inner delaminations (*AM*, *BM*, *CM*) are difficult to detect compared with the edge delaminations (*AU*, *AL*, *BU*, *BL*, *CU*, *CL*). Owing to the boundary condition of inner and edge delaminations, the result is consistent with the physics of the problem. In addition, delaminations that occur near the free end are difficult to detect than the one that occurs near the fixed end. This might be due to the high strain near the fixed end or proximity of sensors to the delaminations near the fixed end or both [60]. From Figure 8b, the difficulty level of delamination detection increases as through-the-thickness interface of delamination is shifted from the mid-plane interface towards the free surfaces. This is because structural stiffness is more severely affected by delamination at the mid-plane than delamination of the same size away from the mid-plane interface.

Contemporary research on delamination assessment (i.e., detection, quantification, and localization) in laminated composites focuses mainly on the use of high-frequency guided-waves (e.g., Lamb waves), acoustic emission/ultrasonic, and mode shape curvatures [61–64]. However, dealing with high-frequency signals may often cause difficulties in terms generation in a specific high-frequency band, high sampling rate, acquisition, storage, a requirement of the controlled experimental environment, and complex signal processing [41,65]. Similarly, acquiring mode shapes experimentally is an expensive and difficult task. On the contrary, low-frequency structural vibration is relatively easier to deal with due to its low sampling rate, small storage requirement, and ease of measurement through piezoelectric sensors. In addition, structural vibration is produced during operation of the component/structure and no input source of excitation is required. However, in most of the published literature, the parameters of structural vibration (e.g., natural frequency, modal damping) have been used for the global assessment (i.e., presence) of delamination [66–69]. The results of the methodology proposed in this work show the possibility to employ low-frequency structural vibration for the global (i.e., detection) and local assessment (i.e., localization, quantification) of delamination in smart composite laminates. In addition, the proposed methodology does not require labor-intensive hand-crafted features and automatically extract discriminative features for the assessment of delaminations in smart composite laminates. The key contribution of the current work is that delamination has been localized by assessing low-frequency structural vibration responses via a convolutional neural network (CNN). Although the large amount of damaged data for the CNN has been produced via numerical simulations, the proposed approach would be extended to real application with limited damaged data through data augmentation and transfer learning in future work.

6. Conclusions

This paper proposed a deep learning framework for the assessment of delamination in smart composite laminates using low-frequency structural vibration responses. The initial assessment of transient responses revealed that multifarious excitations caused more variance in the response behavior than the variance caused by delamination in the structure. A convolutional neural network (CNN) was designed to automatically extract discriminative features from the vibration-based spectrograms and use those features to classify the healthy and delaminated smart composite laminates. The proposed architecture of the CNN showed a training accuracy of 99.9%, a validation accuracy of 97.1%, and a test accuracy of 94.5%. The test confusion chart of the pre-trained CNN revealed that the proposed approach not only distinguished the healthy and delaminated smart composite laminates with 100% accuracy, but also identified the more severe cases of delamination with 95%~100% accuracy. Furthermore, it was found that the detectability of delamination becomes difficult as the in-plane location of delamination is moved from the clamped end of the plate towards the free end and when through-the-thickness interfaces are moved from the mid-plane interface towards the free surfaces. Results of the proposed approach were found to be consistent with the physics of the problem. The proposed approach does not require the labor-intensive process of hand-crafted discriminative features and requires only low-frequency structural vibration response for the global and local assessment of delamination in smart composite laminates. In the future, the proposed approach would be extended to assess multiple damages (e.g., single and multiple delaminations, crack in matrix material, partial debonding of transducer) in cross-paly and angle-ply smart composite laminates. In addition, the proposed deep learning strategy could be used to bridge the gap between simulations, lab scale experiments, and large-scale structures of composite materials via transfer learning or data augmentation.

Author Contributions: Conceptualization, H.S.K. and A.K.; methodology, A.K. and J.K.S.; software, N.Y.K. and W.C.L.; formal analysis, A.K. and J.K.S.; resources, W.C.L.; writing—original draft preparation, A.K. and W.C.L.; writing—review and editing, A.K. and H.S.K.; supervision, H.S.K. All authors have read and agreed to the published version of the manuscript.

Funding: This research was supported by the Basic Science Research Program, through the National Research Foundation of Korea (NRF-2017R1D1A1B03028368 and 2020R1A2C1006613), funded by the Ministry of Education.

Conflicts of Interest: The authors declare no conflict of interest.

Appendix A

$$[L_u] = \begin{bmatrix} 1 & 0 & C_1^k\frac{\partial}{\partial x} + D_1^k\frac{\partial}{\partial y} & A_1^k & B_1^k & H(z-z_k) & \overline{0} & E_1^k\frac{\partial}{\partial x} + F_1^k\frac{\partial}{\partial y} \\ 0 & 1 & C_2^k\frac{\partial}{\partial x} + D_2^k\frac{\partial}{\partial y} & A_2^k & B_2^k & \overline{0} & H(z-z_k) & E_2^k\frac{\partial}{\partial x} + F_2^k\frac{\partial}{\partial y} \\ 0 & 0 & 1 & 0 & 0 & \overline{0} & \overline{0} & H(z-z_k) \end{bmatrix} \tag{A1}$$

$$[L_\varepsilon] = \begin{bmatrix} \frac{\partial}{\partial x} & 0 & C_1^k \frac{\partial^2}{\partial x^2} + D_1^k \frac{\partial^2}{\partial x \partial y} & A_1^k \frac{\partial}{\partial x} & B_1^k \frac{\partial}{\partial x} & H(z-z_j)\frac{\partial}{\partial x} & \bar{0} & E_1^k \frac{\partial^2}{\partial x^2} + F_1^k \frac{\partial^2}{\partial x \partial y} \\ 0 & \frac{\partial}{\partial y} & C_2^k \frac{\partial^2}{\partial x \partial y} + D_2^k \frac{\partial^2}{\partial y^2} & A_2^k \frac{\partial}{\partial y} & B_2^k \frac{\partial}{\partial y} & \bar{0} & H(z-z_j)\frac{\partial}{\partial y} & E_2^k \frac{\partial^2}{\partial x \partial y} + F_2^k \frac{\partial^2}{\partial y^2} \\ 0 & 0 & C_{2,z}^k \frac{\partial}{\partial x} + \left(1 + D_{2,z}^k\right)\frac{\partial}{\partial y} & A_{2,z}^k & B_{2,z}^k & \bar{0} & \bar{0} & E_{2,z}^k \frac{\partial}{\partial x}\left(F_{2,z}^k + H(z-z_j)\right)\frac{\partial}{\partial y} \\ 0 & 0 & \left(C_{1,z}^k + 1\right)\frac{\partial}{\partial x} + D_{1,z}^k \frac{\partial}{\partial y} & A_{1,z}^k & B_{1,z}^k & \bar{0} & \bar{0} & \left[E_{1,z}^k + H(z-z_j)\right]\frac{\partial}{\partial x} + E_{1,z}^k \frac{\partial}{\partial y} \\ \frac{\partial}{\partial y} & \frac{\partial}{\partial x} & C_2^k \frac{\partial^2}{\partial x^2} + \left(C_1^k + D_1^k\right)\frac{\partial^2}{\partial x \partial y} + D_1^k \frac{\partial^2}{\partial y^2} & A_1^k \frac{\partial}{\partial y} + A_2^k \frac{\partial}{\partial x} & B_1^k \frac{\partial}{\partial y} + B_2^k \frac{\partial}{\partial x} & H(z-z_j)\frac{\partial}{\partial y} & H(z-z_j)\frac{\partial}{\partial x} & E_{2,z}^k \frac{\partial^2}{\partial x^2} + \left(E_1^k + F_2^k\right)\frac{\partial^2}{\partial x \partial y} + F_1^k \frac{\partial^2}{\partial y^2} \end{bmatrix} \quad \text{(A2)}$$

$$\left\{L_\phi^p\right\} = \begin{bmatrix} 1 - 4\frac{\left(z - z_0^p\right)^2}{(h^p)^2} & -\left(z - z_0^p\right) + 4\frac{\left(z - z_0^p\right)^3}{(h^p)^2} \end{bmatrix} \quad \text{(A3)}$$

$$\left[L_E^p\right] = \begin{bmatrix} \left(1 - 4\frac{\left(1 - z_0^p\right)^2}{(h^p)^2}\right)\frac{\partial}{\partial x} & \left(4\frac{\left(1 - z_0^p\right)^3}{(h^p)^2} - \left(z - z_0^p\right)\right)\frac{\partial}{\partial x} \\ \left(1 - 4\frac{\left(1 - z_0^p\right)^2}{(h^p)^2}\right)\frac{\partial}{\partial y} & \left(4\frac{\left(1 - z_0^p\right)^3}{(h^p)^2} - \left(z - z_0^p\right)\right)\frac{\partial}{\partial y} \\ -8\frac{\left(1 - z_0^p\right)}{(h^p)^2} & 12\frac{\left(1 - z_0^p\right)^2}{(h^p)^2} - 1 \end{bmatrix} \quad \text{(A4)}$$

$$V_b = 4\overline{\phi}^p \frac{\left(1 - z_0^p\right)^3}{(h^p)^3} \quad \text{(A5)}$$

$$\{F_b\} = \begin{bmatrix} 0 & 0 & 12\overline{\phi}^p \frac{\left(1 - z_0^p\right)^2}{(h^p)^3} \end{bmatrix}^T \quad \text{(A6)}$$

References

1. Elmarakbi, A. *Advanced Composite Materials for Automotive Applications: Structural Integrity and Crashworthiness*; John Wiley & Sons Inc.: Hoboken, NJ, USA, 2013.
2. Chung, D.D. *Composite Materials: Science and Applications*; Springer-Verlag: London, UK, 2010. [CrossRef]
3. Giurgiutiu, V. *Structural Health Monitoring of Aerospace Composites*; Academic Press: London, UK, 2015.
4. Irving, P.E.; Soutis, C. *Polymer Composites in the Aerospace Industry*; Woodhead Publishing: Cambridge, UK, 2019.
5. Mei, H.; Giurgiutiu, V. Guided wave excitation and propagation in damped composite plates. *Struct. Health Monit.* **2019**, *18*, 690–714. [CrossRef]
6. Talreja, R. Manufacturing defects in composites and their effects on performance. In *Polymer Composites in the Aerospace Industry*; Irving, P., Ed.; Woodhead Publishing: Cambridge, UK, 2020; pp. 83–97.
7. Shi, Y.; Swait, T.; Soutis, C. Modelling damage evolution in composite laminates subjected to low velocity impact. *Composite Struct.* **2012**, *94*, 2902–2913. [CrossRef]
8. Khan, A.; Kim, H.S.; Youn, B.D. Modeling and assessment of partially debonded piezoelectric sensor in smart composite laminates. *Int. J. Mech. Sci.* **2017**, *131–132*, 26–36. [CrossRef]
9. Trendafilova, I.; Palazzetti, R.; Zucchelli, A. Delamination assessment in structures made of composites based on general signal correlation. *Int. J. Struct. Stab. Dyn.* **2014**, *14*, 1440022. [CrossRef]
10. Kashtalyan, M.; Soutis, C. The effect of delaminations induced by transverse cracks and splits on stiffness properties of composite laminates. *Compos. Part A Appl. Sci. Manuf.* **2000**, *31*, 107–119. [CrossRef]
11. Wilk, J. Assessing the hazard of delamination propagation in composites using numerical analysis. *Compos. Theory Pract.* **2015**, *15*, 34–38.
12. Haselbach, P.U.; Bitsche, R.D.; Branner, K. The effect of delaminations on local buckling in wind turbine blades. *Renew. Energy* **2016**, *85*, 295–305. [CrossRef]
13. Duchene, P.; Chaki, S.; Ayadi, A.; Krawczak, P. A review of non-destructive techniques used for mechanical damage assessment in polymer composites. *J. Mater. Sci.* **2018**, *53*, 7915–7938. [CrossRef]
14. Balakrishnan, V.S.; Seidlitz, H. Potential repair techniques for automotive composites: A review. *Compos. Part B Eng.* **2018**, *145*, 28–38. [CrossRef]

15. Khan, A.; Kim, H.S. Assessment of sensor debonding failure in system identification of smart composite laminates. *NDT E Int.* **2018**, *93*, 24–33. [CrossRef]
16. Khan, A.; Lee, H.S.; Kim, H.S. Analysis of sensor-debonding failure in active vibration control of smart composite plate. *J. Intell. Mater. Syst. Struct.* **2017**, *28*, 2603–2616. [CrossRef]
17. Soman, R.N.; Majewska, K.; Mieloszyk, M.; Ostachowicz, W. Damage assessment in composite beam using infrared thermography, optical sensors, and terahertz technique. *J. Nondestruct. Eval. Diagn. Progn. Eng. Syst.* **2018**, *1*, 031001. [CrossRef]
18. Cheng, L.; Tian, G.Y. Comparison of nondestructive testing methods on detection of delaminations in composites. *J. Sens.* **2012**, *2012*, 1–7. [CrossRef]
19. Katunin, A.; Dragan, K.; Dziendzikowski, M. Damage identification in aircraft composite structures: A case study using various non-destructive testing techniques. *Compos. Struct.* **2015**, *127*, 1–9. [CrossRef]
20. Jasinien, E.; Raiutis, R.; Voleiis, A.; Vladiauskas, A.; Mitchard, D.; Amos, M. NDT of wind turbine blades using adapted ultrasonic and radiographic techniques. *Insight-Non-Destr. Test. Cond. Monit.* **2009**, *51*, 477–483. [CrossRef]
21. Garcia, C.; Trendafilova, I. Triboelectric sensor as a dual system for impact monitoring and prediction of the damage in composite structures. *Nano Energy* **2019**, *60*, 527–535. [CrossRef]
22. Zhao, G.; Wang, B.; Hao, W.; Luo, Y.; Chen, H. Localization and characterization of delamination in laminates using the local wavenumber method. *Compos. Struct.* **2020**, *238*, 111972. [CrossRef]
23. Grassia, L.; Iannone, M.; Califano, A.; D'Amore, A. Strain based method for monitoring the health state of composite structures. *Compos. Part B Eng.* **2019**, *176*, 107253. [CrossRef]
24. Yang, C.; Oyadiji, S.O. Detection of delamination in composite beams using frequency deviations due to concentrated mass loading. *Compos. Struct.* **2016**, *146*, 1–13. [CrossRef]
25. Mei, H.; Migot, A.; Haider, M.F.; Joseph, R.; Bhuiyan, M.Y.; Giurgiutiu, V. Vibration-Based In-Situ Detection and Quantification of Delamination in Composite Plates. *Sensors* **2019**, *19*, 1734. [CrossRef] [PubMed]
26. Sikdar, S.; Fiborek, P.; Kudela, P.; Banerjee, S.; Ostachowicz, W. Effects of debonding on Lamb wave propagation in a bonded composite structure under variable temperature conditions. *Smart Mater. Struct.* **2018**, *28*, 015021. [CrossRef]
27. Kundu, A.; Sikdar, S.; Eaton, M.; Navaratne, R. A Generic Framework for Application of Machine Learning in Acoustic Emission-Based Damage Identification. In Proceedings of the 13th International Conference on Damage Assessment of Structures, Porto, Portugal, 9–10 July 2019; pp. 244–262. [CrossRef]
28. Khan, A.; Kim, H.S. Assessment of delaminated smart composite laminates via system identification and supervised learning. *Compos. Struct.* **2018**, *206*, 354–362. [CrossRef]
29. Chen, D.-M.; Xu, Y.F.; Zhu, W.D. A Comprehensive Study on Detection of Hidden Delamination Damage in a Composite Plate Using Curvatures of Operating Deflection Shapes. *J. Nondestruct. Eval.* **2019**, *38*, 54. [CrossRef]
30. Yelve, N.P.; Mitra, M.; Mujumdar, P.M. Detection of delamination in composite laminates using Lamb wave based nonlinear method. *Compos. Struct.* **2017**, *159*, 257–266. [CrossRef]
31. Feng, B.; Ribeiro, A.L.; Ramos, H.G. Using guided ultrasonic wave inspection to quantify the length of delaminations in composite laminates. In *AIP Conference Proceedings*; AIP Publishing LLC: Melville, NY, USA, 2018.
32. Dafydd, I.; Sharif Khodaei, Z. Analysis of barely visible impact damage severity with ultrasonic guided Lamb waves. *Struct. Health Monit.* **2019**. [CrossRef]
33. Li, G.; Neerukatti, R.K.; Chattopadhyay, A. Ultrasonic guided wave propagation in composites including damage using high-fidelity local interaction simulation. *J. Intell. Mater. Syst. Struct.* **2018**, *29*, 969–985. [CrossRef]
34. Huang, B.; Koh, B.-H.; Kim, H.S. PCA-based damage classification of delaminated smart composite structures using improved layerwise theory. *Comput. Struct.* **2014**, *141*, 26–35. [CrossRef]
35. Minak, G.; Palazzetti, R.; Trendafilova, I.; Zucchelli, A. Localization of a delamination and estimation of its length in a composite laminate beam by the VSHM and pattern recognition methods. *Mech. Compos. Mater.* **2010**, *46*, 387–394. [CrossRef]
36. Montalvao, D. A Review of Vibration-based Structural Health Monitoring with Special Emphasis on Composite Materials. *Shock Vib. Dig.* **2006**, *38*, 295–324. [CrossRef]

37. Yang, B.; Xuan, F.-Z.; Chen, S.; Zhou, S.; Gao, Y.; Xiao, B. Damage localization and identification in WGF/epoxy composite laminates by using Lamb waves: Experiment and simulation. *Compos. Struct.* **2017**, *165*, 138–147. [CrossRef]
38. Prasad, S.M.; Balasubramaniam, K.; Krishnamurthy, C.V. Structural health monitoring of composite structures using Lamb wave tomography. *Smart Mater. Struct.* **2004**, *13*, N73. [CrossRef]
39. Kessler, S.S.; Spearing, S.M.; Atalla, M.J. In-situ damage detection of composites structures using Lamb wave methods. In Proceedings of the First European Workshop on Structural Health Monitoring, Paris, France, 10–12 July 2002; pp. 374–381.
40. Abbas, M.; Shafiee, M. Structural health monitoring (SHM) and determination of surface defects in large metallic structures using ultrasonic guided waves. *Sensors* **2018**, *18*, 3958. [CrossRef] [PubMed]
41. Su, Z.; Ye, L.; Lu, Y. Guided Lamb waves for identification of damage in composite structures: A review. *J. Sound Vib.* **2006**, *295*, 753–780. [CrossRef]
42. Bogue, R. Smart materials: A review of capabilities and applications. *Assem. Autom.* **2014**, *34*, 16–22. [CrossRef]
43. Leckey, C.A.; Wheeler, K.R.; Hafiychuk, V.N.; Hafiychuk, H.; Timuçin, D.A. Simulation of guided-wave ultrasound propagation in composite laminates: Benchmark comparisons of numerical codes and experiment. *Ultrasonics* **2018**, *84*, 187–200. [CrossRef]
44. Memmolo, V.; Monaco, E.; Boffa, N.D.; Maio, L.; Ricci, F. Guided wave propagation and scattering for structural health monitoring of stiffened composites. *Compos. Struct.* **2018**, *184*, 568–580. [CrossRef]
45. Sha, G.; Radzienski, M.; Soman, R.; Cao, M.; Ostachowicz, W.; Xu, W. Multiple damage detection in laminated composite beams by data fusion of Teager energy operator-wavelet transform mode shapes. *Compos. Struct.* **2020**, *235*, 111798. [CrossRef]
46. Raut, N.P.; Kolekar, A.B.; Gombi, S.L. Methods of damage detection on composites under low velocity impact. *Mater. Today Proc.* **2020**. [CrossRef]
47. Tiachacht, S.; Slimani, M.; Khatir, S.; Behtani, A.; Mansouri, L.; Bouazzouni, A.; Wahab, M.A. Damage Assessment of Laminated Composite Plates Using a Modified Cornwell Indicator. In Proceedings of the 13th International Conference on Damage Assessment of Structures, Porto, Portugal, 9–10 July 2019; pp. 853–862. [CrossRef]
48. Dong, S.; Yuan, M.; Wang, Q.; Liang, Z. A Modified Empirical Wavelet Transform for Acoustic Emission Signal Decomposition in Structural Health Monitoring. *Sensors* **2018**, *18*, 1645. [CrossRef]
49. Godin, N.; Reynaud, P.; Fantozzi, G. Challenges and Limitations in the Identification of Acoustic Emission Signature of Damage Mechanisms in Composites Materials. *Appl. Sci.* **2018**, *8*, 1267. [CrossRef]
50. Kim, H.S.; Chattopadhyay, A.; Ghoshal, A. Dynamic Analysis of Composite Laminates with Multiple Delamination Using Improved Layerwise Theory. *AIAA J.* **2003**, *41*, 1771–1779. [CrossRef]
51. Kim, H.S.; Ghoshal, A.; Chattopadhyay, A.; Prosser, W.H. Development of Embedded Sensor Models in Composite Laminates for Structural Health Monitoring. *J. Reinf. Plast. Compos.* **2004**, *23*, 1207–1240. [CrossRef]
52. Huang, B.; Kim, H.S.; Yoon, G.H. Modeling of a partially debonded piezoelectric actuator in smart composite laminates. *Smart Mater. Struct.* **2015**, *24*, 075013. [CrossRef]
53. Shaheen, F.; Verma, B.; Asafuddoula, M. Impact of Automatic Feature Extraction in Deep Learning Architecture. In Proceedings of the 2016 International Conference on Digital Image Computing: Techniques and Applications (DICTA), Queensland, Australia, 30 November–2 December 2016; pp. 1–8.
54. Cerna, M.; Harvey, A.F. The Fundamentals of FFT-Based Signal Analysis and Measurement; Application Note 041, National Instruments. 2000. Available online: http://www.ni.com/white-paper/4278/en/ (accessed on 18 April 2020).
55. Zhang, X.; Zou, Y.; Shi, W. Dilated convolution neural network with LeakyReLU for environmental sound classification. In Proceedings of the 2017 22nd International Conference on Digital Signal Processing (DSP), London, UK, 23–25 August 2017; pp. 1–5.
56. Wang, S.-H.; Phillips, P.; Sui, Y.; Liu, B.; Yang, M.; Cheng, H. Classification of Alzheimer's Disease Based on Eight-Layer Convolutional Neural Network with Leaky Rectified Linear Unit and Max Pooling. *J. Med. Syst.* **2018**, *42*, 85. [CrossRef]

57. Habibi Aghdam, H.; Jahani Heravi, E. (Eds.) Convolutional Neural Networks. In *Guide to Convolutional Neural Networks: A Practical Application to Traffic-Sign Detection and Classification*; Springer International Publishing: Cham, Switzerland, 2017; pp. 85–130. ISBN 978-3-319-57550-6.
58. Aggarwal, C.C. Convolutional Neural Networks. In *Neural Networks and Deep Learning: A Textbook*; Springer International Publishing: Cham, Switzerland, 2018; pp. 315–371. ISBN 978-3-319-94463-0.
59. Sultan, R.; Guirguis, S.; Younes, M.; El-Soaly, E. Effect of the Thickness-wise Location Delamination on Natural Frequency for Laminate Composite. *Int. J. Appl. Eng. Res.* **2013**, *8*, 157–170.
60. Kim, H.S.; Kim, J.; Choi, S.-B.; Ghoshal, A.; Chattopadhyay, A. Modal-Strain-Based Damage Index of Laminated Composite Structures Using Smooth Transition of Displacements. *AIAA J.* **2007**, *45*, 2972–2978. [CrossRef]
61. Wang, K.; Liu, M.; Cao, W.; Yang, W.; Su, Z.; Cui, F. Detection and sizing of disbond in multilayer bonded structure using modally selective guided wave. *Struct. Health Monit.* **2019**. [CrossRef]
62. Mei, H.; James, R.; Haider, M.F.; Giurgiutiu, V. Multimode Guided Wave Detection for Various Composite Damage Types. *Appl. Sci.* **2020**, *10*, 484. [CrossRef]
63. Barman, S.K.; Jebieshia, T.R.; Tiwari, P.; Maiti, D.K.; Maity, D. Two-Stage Inverse Method to Detect Delamination in Composite Beam Using Vibration Responses. *AIAA J.* **2019**, *57*, 1312–1322. [CrossRef]
64. Gómez Muñoz, C.Q.; García Marquez, F.P.; Hernandez Crespo, B.; Makaya, K. Structural health monitoring for delamination detection and location in wind turbine blades employing guided waves. *Wind Energy* **2019**, *22*, 698–711. [CrossRef]
65. Khan, A.; Kim, N.; Shin, J.K.; Kim, H.S.; Youn, B.D. Damage assessment of smart composite structures via machine learning: A review. *JMST Adv.* **2019**, *1*, 107–124. [CrossRef]
66. Della, C.N.; Shu, D. Vibration of Delaminated Composite Laminates: A Review. *Appl. Mech. Rev.* **2007**, *60*, 1–20. [CrossRef]
67. Dahak, M.; Touat, N.; Kharoubi, M. Damage detection in beam through change in measured frequency and undamaged curvature mode shape. *Inverse Probl. Sci. Eng.* **2019**, *27*, 89–114. [CrossRef]
68. Ullah, I.; Sinha, J.K.; Pinkerton, A. Vibration-Based Delamination Detection in a Composite Plate. *Mech. Adv. Mater. Struct.* **2013**, *20*, 536–551. [CrossRef]
69. Huang, B.; Kim, H.S. Frequency response analysis of a delaminated smart composite plate. *J. Intell. Mater. Syst. Struct.* **2015**, *26*, 1091–1102. [CrossRef]

sensors

MDPI

Article

Development of a Flex and Stretchy Conductive Cotton Fabric Via Flat Screen Printing of PEDOT:PSS/PDMS Conductive Polymer Composite

Granch Berhe Tseghai [1,2,3,*], Benny Malengier [1], Kinde Anlay Fante [3], Abreha Bayrau Nigusse [1,2] and Lieva Van Langenhove [1]

1 Department of Materials, Textiles and Chemical Engineering, Ghent University, 9000 Gent, Belgium; Benny.malengier@UGent.be (B.M.); AbrehaBayrau.BayrauNigusse@UGent.be (A.B.N.); Lieva.VanLangenhove@UGent.be (L.V.L.)

2 Ethiopian Institute of Textile and Fashion Technology, Bahir Dar University, 6000 Bahir Dar, Ethiopia

3 Jimma Institute of Technology, Jimma University, Jimma, Ethiopia; kinde.anlay@ju.edu.et

* Correspondence: GranchBerhe.Tseghai@UGent.be; Tel.: +32465570635

Received: 27 February 2020; Accepted: 18 March 2020; Published: 20 March 2020

Abstract: In this work, we have successfully produced a conductive and stretchable knitted cotton fabric by screen printing of poly(3,4-ethylenedioxythiophene) polystyrene sulfonate (PEDOT:PSS) and poly(dimethylsiloxane-b-ethylene oxide)(PDMS-b-PEO) conductive polymer composite. It was observed that the mechanical and electrical properties highly depend on the proportion of the polymers, which opens a new window to produce PEDOT:PSS-based conductive fabric with distinctive properties for different application areas. The bending length analysis proved that the flexural rigidity was lower with higher PDMS-b-PEO to PEDOT:PSS ratio while tensile strength was increased. The SEM test showed that the smoothness of the fabric was better when PDMS-b-PEO is added compared to PEDOT:PSS alone. Fabrics with electrical resistance from 24.8 to 90.8 kΩ/sq have been obtained by varying the PDMS-b-PEO to PEDOT:PSS ratio. Moreover, the resistance increased with extension and washing. However, the change in surface resistance drops linearly at higher PDMS-b-PEO to PEDOT:PSS ratio. The conductive fabrics were used to construct textile-based strain, moisture and biopotential sensors depending upon their respective surface resistance.

Keywords: conductive polymer composite; PEDOT:PSS; flexible electronics; wearable application

1. Introduction

Poly(3,4-ethylenedioxythiophene) polystyrene sulfonate (PEDOT:PSS) conductive polymer is well-known for its high conductivity and applications in conductive synthetic textiles. It has been used with encouraging results as electrodes for flexible electronics. Unfortunately the use of PEDOT:PSS is currently constrained by its brittleness and limited processability. As a result many researchers have been trying preparing PEDOT:PSS-based conductive polymer composites. For instance, PEDOT:PSS and graphene oxide (GO) as an efficient alternative structure for indium tin oxide (ITO) in organic photovoltaics [1], ITO-PEDOT:PSS/poly(3-hexylthiophene):phenyl-C61-butyric acid methyl ester/Al [2], poly (vinyl alcohol) (PVA)–PEDOT:PSS blend filled with synthesized GO and reduced GO by solvent casting technique [3], PEDOT:PSS@polyurethane nonwovens by electrospinning and dip-coating [4], super paramagnetic PEDOT/magnetite nano particles [5], GO/glucose/PEDOT:PSS super capacitor [6], graphene and poly(3,4-ethylenedioxy thiophene)–poly(styrenesulfonate) (G-PEDOT:PSS) [7] an electroactive bacterium, Shewanellaoneidensis MR-1, inside a conductive three-dimensional PEDOT:PSS matrix [8], thermoelectric PEDOT:PSS with polyurethane [9], UV-ozone treated GO/PEDOT:PSS [10], graphenenano-platelets with PEDOT:PSS solutions to produce conductive, breathable, and light-weight mercerized cotton fabrics by spray coating [11]. PVA, phosphoric acid, PEDOT:PSS, and silver flakes [12], 3D graphene–PEDOT:PSS skeleton with poly(dimethylsiloxane) (PDMS) [13], and single-walled carbon nanotubes/PEDOT:PSS coated Tenano-rod composite films [14] have all been successfully developed. Though promising results have been found, a truly textile-based conductive device with adequate flexibility and stretchability without intensive impact on the bulk property of the textile fabric is still required.

On the other hand highly flexible and biocompatible conductive polymer composites are required. Because it is biocompatible, transparent, gas permeable, and economical, PDMS is widely used in medical research and technology, and there are a wide array of manufacturing techniques used for forming PDMS including soft-lithography and its derivatives, molding, dip casting, spin coating and many others. Besides, PDMS has a dielectric constant and a tunable elasticity which make it suitable for flexible electronic sensors [15]. As a result, PDMS-based flexible composites have been studied by many researchers. For example, high strain biocompatible PDMS-based conductive graphene and multi-walled carbon nanotube as a nano-composite strain sensors [16],electrically conductive PDMS-grafted carbon nanotubes-reinforced silicone elastomer [17], enhanced conductivity behavior of PDMS hybrid composites containing exfoliated graphite nano-platelets and carbon nano-tubes [18], high electro-conductive PDMS/short carbon fiber binary composites with electrical conductivity of 1.67×10^2 S/m [19], conductive elastomers based on multi-walled carbon nanotubes in PDMS with up to 0.01 S/cm conductivity [20], silver nano-wire network embedded in PDMS as a stretchable, transparent, and conductive substrate with 15 Ohm/sq [20], and stretchable electronics based on Ag-PDMS PCB (Printed circuit board) with a typical resistance of 2 Ohms/cm [21], have been reported.

It was noticed that the common limitations of the conductive polymer composite-based conductive textiles reported in many works of the literature are that they possess inadequate flexibility, stretchability, and biocompatibility. Moreover, technical and scientific experimental evidence about the effect the conductive polymer composite has on the textile bulk properties like flexural rigidity, tensile strength, and extension at break, were not reported, which does not allow to determine if the fabrics still remain a true textile or if they lost their texture.

Therefore, another approach is introduced in this work to produce a PEDOT:PSS Clevios PH 1000 (Figure 1a) and PDMS-b-PEO (Figure 1b) conductive polymer composite-based fabric. As fabric, we selected knitted cotton, as cotton fabric is available everywhere and making cotton conductive will make access to conductive fabric easier. In addition, this work contains an in-depth study on the properties of the conductive textile, not only from an electronically point of view, but also as a textile material. The effect of the conductive polymer composite on bending length, flexural rigidity, tensile strength, and extension at break, thickness, add-on and weight has been studied.

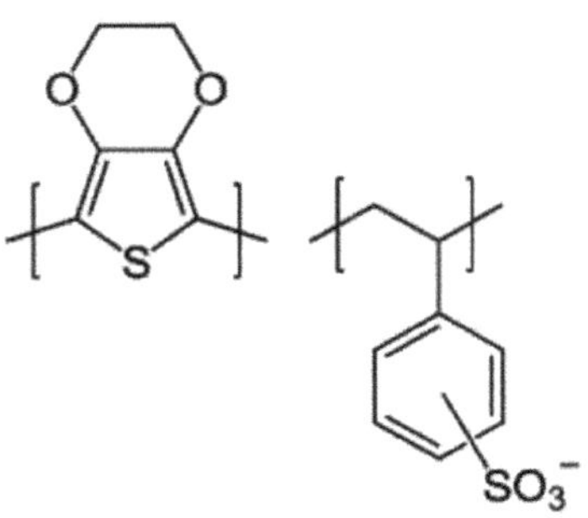

(a)

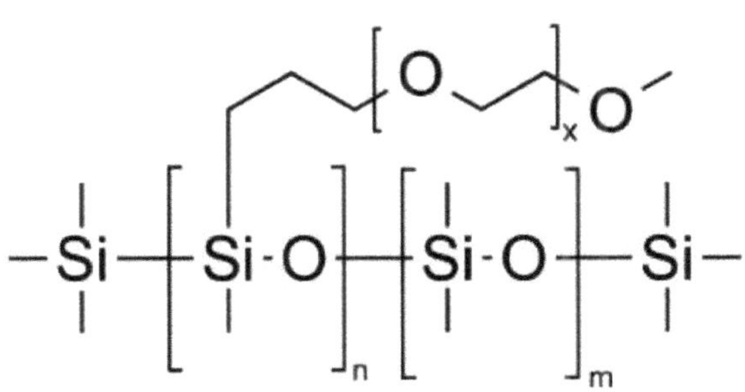

(b)

Figure 1. Chemical structure: (**a**) polystyrene sulfonate (PEDOT:PSS); (**b**) poly(dimethylsiloxane-b-ethylene oxide) (PDMS-b-PEO).

2. Materials and Methods

2.1. Material and Chemicals

The conductive textile fabric produced consists of two parts: a water repellent textile substrate and an electro-conductive polymer composite. A weft knitted cotton fabric (S0) with 140 GSM and 0.5 mm thickness obtained from UGent, MaTCh laboratory was used as the textile fabric. Nano Spray water repellent (WR) for textile obtained from Lab@Home, Netherlands was utilized. PEDOT:PSS PH1000 Clevious conductive polymer obtained from Ossila Ltd. (Sheffield, UK) and poly(dimethylsiloxane-b-ethylene oxide) methyl terminated (PDMS-b-EO) obtained from Polyscience, Inc. (Warrington, UK) were used to produce the conductive polymer composite. The sheet resistance of each conductive polymer composite (PEDOT:PSS/PDMS-b-PEO) screen-printed fabric with different dimensions (sample sizes) was measured by a two point-method.

2.2. Methods

2.2.1. Fabric Water Repellency Pre-Treatment

Before applying the PEDOT:PSS/PDMS-b-PEO conductive polymer composite (CPC), the fabric was pre-treated with WR to impart a hydrophobic effect on the fabric surface, and therefore prevent absorption of CPC into the fabric structure giving a confined CPC distribution within the required area. 3% owf (own weight of fabric) WR was gently sprayed on the fabric surface and then dried at 80 °C for 3 min. A drop test with water and PEDOT:PSS performed on the fabric showed an effective hydrophobicity as shown in Figure 2.

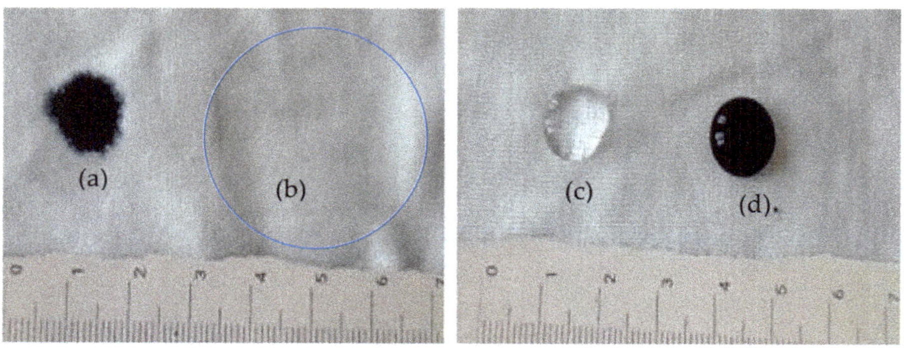

Figure 2. Effect of WR treatment on knitted cotton fabric: (**a**) PEDOT:PSS polymer dispersion on untreated fabric (S0); (**b**) drops of water on untreated fabric (S0); (**c**) drops of water on water repellent (WR)-treated fabric (S1); (**d**) PEDOT:PSS polymer dispersion on water repellent (WR)-treated fabric.

2.2.2. Flat Screen Printing

Flat screen printing was preferred for this work because attaining the required conductive polymer composite paste viscosity and transferring it to the substrate by this technique is straightforward. Other possible techniques that could be explored are transfer printing [22], while also mechanics designs via transfer printing may provide another effective route toward stretchable conductive fabric [23].

Different amounts of PDMS-PEO was manually mixed with a fixed amount of PEDOT:PSS (4 ml) for 3 min at room temperature. A mixing rod was used to prepare a homogenous blend of the two polymers. It was observed that stirring of the polymers showed a shearing property, the mixture solution was converted into a thicker paste which is convenient for screen printing. As a result, any thickening agent like present in conventional screen printing was not employed. Then the paste was applied to the surface of WR-treated cotton fabric (S1) via flat screen printing. A gentle agitation of the PEDOT:PSS and PDMS-b-PEO mix formed a thick paste and then the paste was simply swept over the flat screen mesh by a mini squeegee to improve the distribution. The overall printing process for this work is schematically represented in Figure 3.

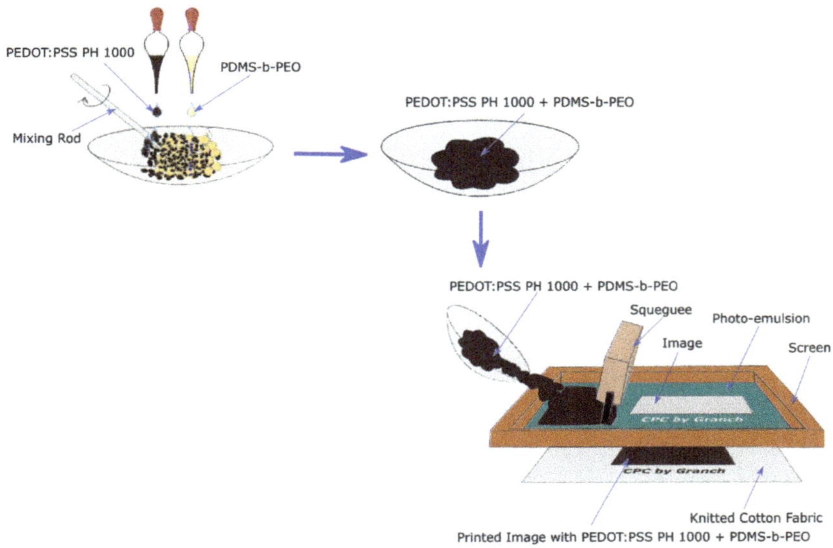

Figure 3. Screen printing of conductive polymer composite on knitted cotton fabric.

After the printing process, drying was performed in an oven at 70 °C for 5 min and cured at 150 °C for 5 min. The printed samples were thoroughly washed with distilled water. An example of the actual conductive knitted fabric produced is shown Figure 4.

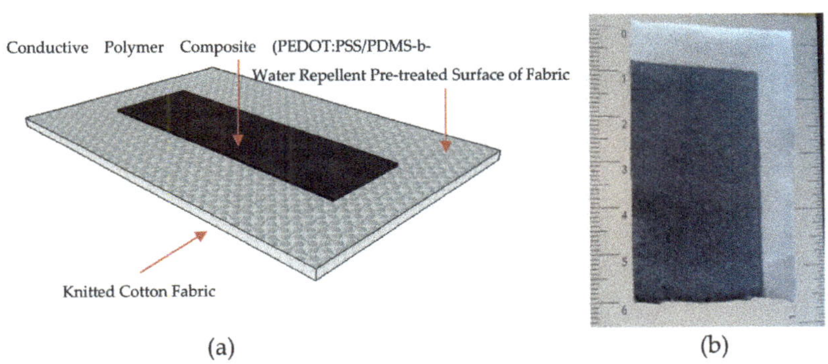

Figure 4. (**a**) Schematic design of the conductive textile fabric; (**b**) actual conductive textile fabric.

To study the effect of PDMS-b-PEO on surface resistance, ten different ratios of PDMS-b-PEO to PEDOT:PSS (Table 1) were prepared and applied on the previously WR-treated knitted fabric via screen printing over an area of 62.5 cm^2 (12.5 cm × 5 cm). The solid add-ons of the PEDOT:PSS/PDMS-b-PEO conductive polymer composite on the textile fabric were calculated using Equation (1).

$$w = (W_a - W_b)/A, \quad (1)$$

where, W_a = weight of fabric after printing in g and W_b = weight of fabric before printing in g, A = coated surface area in cm^2 and w = solid add on per coated surface area (A) in g/cm^2.

Table 1. Mix ratios of the conductive polymer composite and weight before printing.

Sample	r [%]	V_c [ml]	V_e [ml]	W_b [g]	A [cm^2]
r00	0	4	0	1.655	5 × 12.5
r10	10	4	0.4	1.653	5 × 12.5
r20	20	4	0.8	1.629	5 × 12.5
r30	30	4	1.2	1.608	5 × 12.5
r40	40	4	1.6	1.578	5 × 12.5
r50	50	4	2	1.645	5 × 12.5
r60	60	4	2.4	1.678	5 × 12.5
r70	70	4	2.8	1.661	5 × 12.5
r80	80	4	3.2	1.622	5 × 12.5
r90	90	4	3.6	1.628	5 × 12.5

The respective solid add-ons (w) for each percentage volume mix ratio of PDMS-b-PEO to PEDOT:PSS are provided in Table 1. The percentage volume mix ratio was calculated using Equation (2).

$$r = (V_e/V_c) \times 100, \quad (2)$$

where, V_e = volume of PDMS-b-PEO in ml, V_c = volume of PEDOT:PSS in ml, and r = percentage volume mix ratio of PDMS-b-PEO to PEDOT:PSS in %.

2.2.3. Mechanical Characterization

Thickness: The thickness was determined using ISO 5084:1996(E) (determination of thickness of textile and textile products) using "Mitutoyo Digimatic Indicator."

Bending Analysis: The bending length was measured according to the test method BS 3356:1990using a bending meter. Using the appropriate mean value, we calculated the flexural rigidity G, in mg cm, separately for the warp and weft directions by Equation (3).

$$G = 0.1MC^3, \quad (3)$$

where, G = flexural rigidity (mg cm), C = bending length (cm), M = mass/area of the specimen (g/m^2).

Tensile Strength and Elongation at Break: The strength and elongation at break were tested using INSTRON universal strength tester. A tensile test according to ISO 13934-1 was used.

SEM analysis of CPC: The surface topology, cracks, holes, and appearance of yarns within the fabric before and after coatings were studied using FEI Quanta 200 FFE-SEM. Images were taken with an accelerating voltage of 20 kV. The non-conductive samples were prepared prior to analysis by applying a gold coating using Balzers Union SKD 030 sputter coater.

2.2.4. Electrical Characterization

The sheet resistance of all samples was measured by using two-point methods. The samples were placed in a 3-7/8" MaxSteel Light Duty Drill Vise 83070 Stanley Hand Tool on both ends to make them stable for measurement as presented in Figure 5. The effect of stretching on sheet resistance has been studied from 0 to 35% elongation. The effect of repeated stretching on the sheet resistance has been studied by stretching the samples to their infliction point within five seconds, releasing from stretch and measuring the sheet resistance after 5 s.The stretching and releasing have been continued until the sheet resistance reached an infinite value.

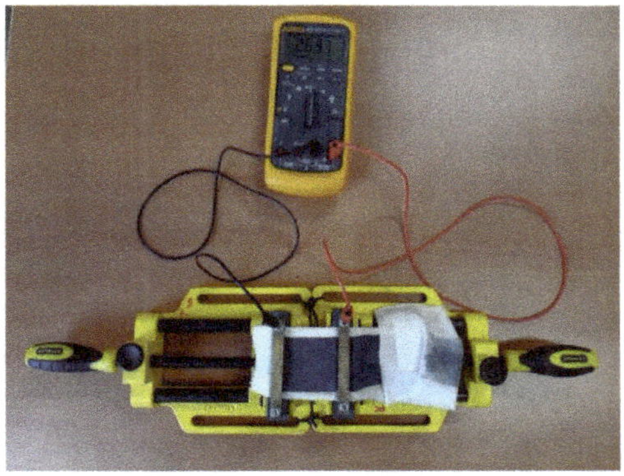

Figure 5. Two-point method resistance measurements set-up.

In addition, we have constructed a 2.5 × 5 cm strain and moisture electrode from sample r60 as a demonstrator. We have also developed a PEDOT:PSS/PDMS-b-PEO (4:1) coated cotton fabric with a 332.5 Ω sheet resistance by increasing the add-on to 0.013 g/cm^2 and tested this for electrocardiography (ECG) and electroencephalography (EEG) electrodes.

To study the sensing stability of the conductive polymer composite-treated fabric an Arduino Nano set-up for this particular purpose with the circuitry in Figure 6 was used. One end of the PEDOT:PSS/PDMS-b-PEO CPC-treated fabric was connected to 5V of Arduino Nano input and the other was connected to 1MΩpull-down resistor and analog-to-digital converter (ADC) input. An IDE program suitable for reading out dynamically the resistance of the CPC was written. Then, the percentage change in resistance at its respective change in elongation from 0 to 35% was calculated using Equation (4).

$$\Delta R = 100 \times (R_f - R_i)/R_i, \tag{4}$$

where, ΔR = percentage change in resistance (%), R_i = initial resistance (kΩ), R_f = final resistance (kΩ).

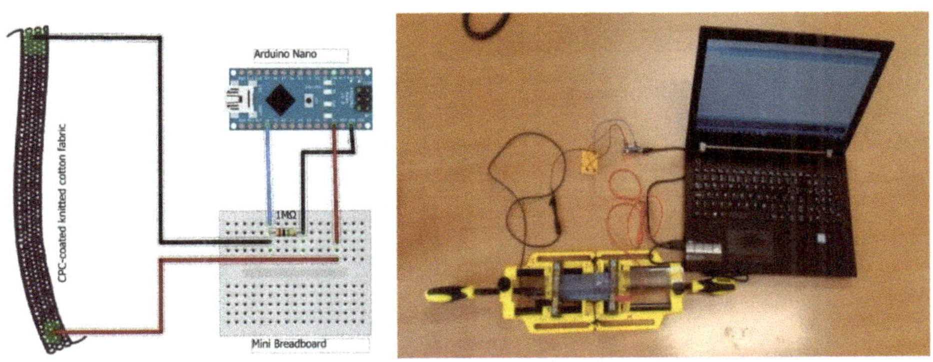

(a) (b)

Figure 6. (**a**) Arduino Nano electronic circuitry design; (**b**) resistance measurements set-up.

In addition, the effect of washing on the surface resistance was measured. PEDOT:PSS/PDMS-b-PEO-coated knitted cotton fabric specimens of 4 × 4 cm were sandwiched between cotton fabric and polyester fabric and then sewn along all the four sides to form a composite specimen. Washing solution containing 5 g/l soap was taken into the launder-o-meter with a liquor ratio of 1:40. The specimen was treated for 30 min at 30 °C at a speed of 40 revolutions per minute. The specimen was removed and rinsed in cold water. The stitch was opened on three sides and dried in shadow. The surface resistance after washing was measured using a multi-meter and the two-point method. Finally, the percentage change in surface resistance was calculated using Equation (4).

3. Result and Discussion

3.1. Mechanical Property Analysis

3.1.1. Solid Add-on

The solid add-on (w) of the conductive polymer composite on the knitted cotton fabric obviously increased with the increase in the amount of PDMS-b-PEO percentage mix ratio as represented in Table 2. This increase in add-on due to the increase in the amount of PDMS-b-PEO may raise the sheet resistance as the conductive component could probably be trapped inside the PDMS-b-PEO elastomer.

Table 2. Solid add-on of coated samples.

Sample	V_c [ml]	V_e [ml]	W_b [g]	W_a [g]	A [cm^2]	w [g/cm^2]
r00	4	0	1.655	1.913	5 × 12.5	0.0041
r10	4	0.4	1.653	2.142	5 × 12.5	0.0078
r20	4	0.8	1.629	2.344	5 × 12.5	0.0114
r30	4	1.2	1.608	2.401	5 × 12.5	0.0127
r40	4	1.6	1.578	2.462	5 × 12.5	0.0141
r50	4	2	1.645	2.44	5 × 12.5	0.0144
r60	4	2.4	1.678	2.646	5 × 12.5	0.0155
r70	4	2.8	1.661	2.671	5 × 12.5	0.0162
r80	4	3.2	1.622	2.688	5 × 12.5	0.0171
r90	4	3.6	1.628	2.733	5 × 12.5	0.0177

3.1.2. Thickness Analysis

From Figure 7, the thickness of the fabric increased by 0.03 mm when WR is applied. After the conductive polymer composite has been applied, the thickness has moreover increased by less than or equal to 0.14 mm until a percentage mix ratio of 60% and continues with a thickness equal to the sample r00 until the mix ratio is 90% (sample r90). This shows that the presence of PDMS-b-PEO has less effect on the thickness than when PEDOT:PSS alone (sample r00) was utilized. This could have a positive contribution to the flexibility of the fabric.

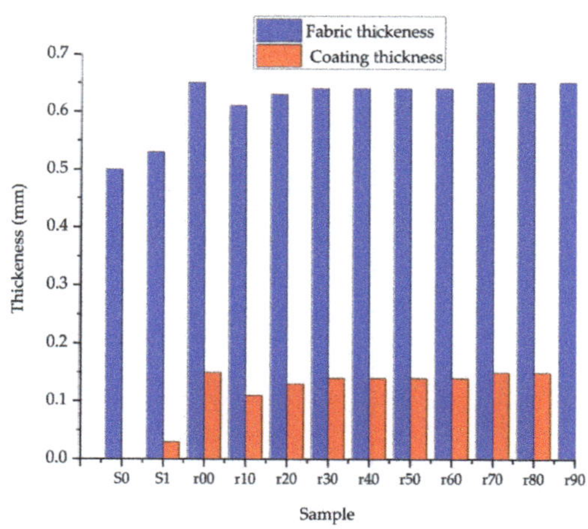

Figure 7. Effect of PEDOT:PSS/PDMS-bb-PEO on thickness.

3.1.3. Bending Length Analysis

In Table 3, we presented the measured values of bending length and respective calculated flexural rigidity according to Equation (3). The fabric became stiffer when treated with a water-repelling agent, as evidenced by an increase in bending length of 29%. The fabric became again stiffer when the CPC was applied, with double the bending length of the r00 sample over the S0 sample (base cotton fabric). Moreover, this increase in bending length was more intensive when PEDOT:PSS alone was used. For instance, the bending length of the PEDOT:PSS-treated fabric decreased on an average by 11% when r was 50% (sample r50) and the flexural rigidity reduced by 13%.

Table 3. Bending length and flexural rigidly results.

Fabric	Weight (g/m^2)	Bending Length (cm)					Flexural Rigidity (mg cm)				
		Wale Direction		Course Direction		Average	Wale Direction		Course Direction		Average
		Face	Back	Face	Back		Face	Back	Face	Back	
S0	140.0	1.3	1.2	1.3	1.7	1.4	27.3	24.2	27.3	71.2	34.8
S1	146.0	1.8	1.7	1.7	1.9	1.8	78.2	65.6	71.7	100.1	78.2
r00	176.0	3.0	2.9	2.9	2.5	2.8	475.2	407.4	407.4	262.0	382.2
r10	191.0	2.9	2.8	2.8	2.4	2.7	456.3	397.2	428.3	247.9	375.9
r20	199.0	2.9	2.7	2.7	2.3	2.7	460.7	396.1	409.4	245.3	371.4
r30	205.0	2.8	2.7	2.7	2.2	2.6	454.9	394.6	408.0	230.4	364.5
r40	209.0	2.8	2.7	2.6	2.2	2.6	453.9	388.9	371.6	219.5	350.6
r50	213.0	2.7	2.6	2.6	2.1	2.5	433.4	383.1	353.2	200.1	333.8
r60	216.0	2.7	2.6	2.5	2.1	2.5	415.8	370.9	341.6	191.6	321.6
r70	221.0	2.6	2.5	2.5	2.0	2.4	392.9	349.5	341.2	179.5	307.4
r80	224.0	2.5	2.5	2.5	1.9	2.3	362.8	337.6	333.5	153.6	287.0
r90	228.0	2.5	2.5	2.4	1.9	2.3	356.3	335.3	315.2	149.1	279.2

3.1.4. Tensile Strength

From the load-elongation curve in Figure 8, both the tensile strength and strain at break of the fabric reduced from 72.2 to 68.1 N and 115.3 to 112.3% because of the water repellent treatment. This could be due to the effect of the water repellent that might cause a very slight degradation during drying and curing, but is overall a minor influence on the textile properties.

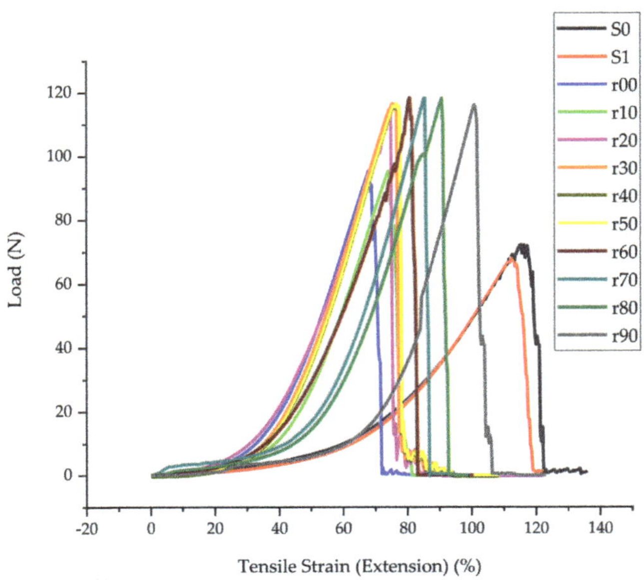

Figure 8. Load-elongation curve.

The application of only PEDOT:PSS (r00) increased the tensile strength in line with a stiffer fabric as found in the bending testing, but the strain was greatly decreased. This indicates that the PEDOT:PSS on the surface adds to the strength of the fabric making it stiffer, but reduces the stretch. The addition of PDMS-PEO to the fabric gives again higher tensile strength and allows again higher strain. Further increase of PDMS-PEO ratio has little influence on the strength, but does further increase the strain. For instance, a shift of tensile strength from 78.2 to 115.8 N and a strain 69.7 to 77 % at break was observed when r shifts from 0% to 50%. In general, the tensile strength becomes constant at around 118 N. Therefore, it is rational to say the new conductive polymer composite gives smaller Young's modulus which is important for the e-textile application.

3.1.5. SEM Characteristics of the Conductive Polymer Composite

The SEM results (Figure 9) showed that the yarn loop interstices in the fabric were coved by the addition of PEDOT:PSS/PDMS-b-PEO conductive polymer composite. Protruding loops were also observed in the base fabric but not after the printing. As a result, the coated fabrics are smoother than uncoated fabrics. The presence of PDMS-PEO further improved the smoothness and the coverage of the yarn loop interstices. The surface of the PEDOT:PSS-treated fabric looks more glassy when PDMS-b-PEO was added.

(a) **(b)** **(c)**

Figure 9. SEM: (**a**) WR treated; (**b**) PEDOT:PSS coated; (**c**) PEDOT:PSS-PDMS:PEO coated.

3.2. *Electrical Characteristics Analysis*

3.2.1. Effect of PDMS-b-PEO to PEDOT:PSS Ratio on Resistance

For all the samples, the resistance increases with increase in the surface area when the length was increased by keeping the width constant. Therefore, it is obvious that resistance drops with increasing concentration of the non-conductive polymer and the surface area. The effect of the PDMS-b-PEO to PEDOT:PSS ratio on the surface resistance is shown in Figure 10.

In all the samples, it was observed that the reproducibility was fairly good but still needs improvement. This could arise from the homogeneity of the polymer blend, uneven distribution of the conductive polymer composite during screen printing, and the uneven property of the textile fabric at its different portions.

As the obtained resistance covers a big range from 6 to 468.4 kΩ (Figure 10b) depending on the concentration of PDMS-b-PEO and the surface area of the treated fabric, they can be used as a textile-based electrode for different sensors such as strain, moisture, biopotential (EEG, ECG), interconnection, energy storage, and other applications. For instance, in the strain and moisture sensor electrode demonstration of sample r60, it showed a linear increase in resistance during the stretch to its infliction point. The resistance also dropped up to 132.5% moisture regain, while above 132.5%, the resistance rapidly increased, which may be due to swelling of the PEDOT:PSS at higher moisture regain. Moreover, the ECG and EEG electrodes constructed from the PEDOT:PSS/PDMS-b-PEO (4:1) coated cotton fabric (332.5 Ω) at an add-on of 0.013g/cm^2, showed good qualities of collected ECG and EEG waveforms. The strain, moisture, ECG and EEG responses from the PEDOT:PSS/PDMS-b-PEO-coated cotton fabric electrodes are shown in Figure 11.

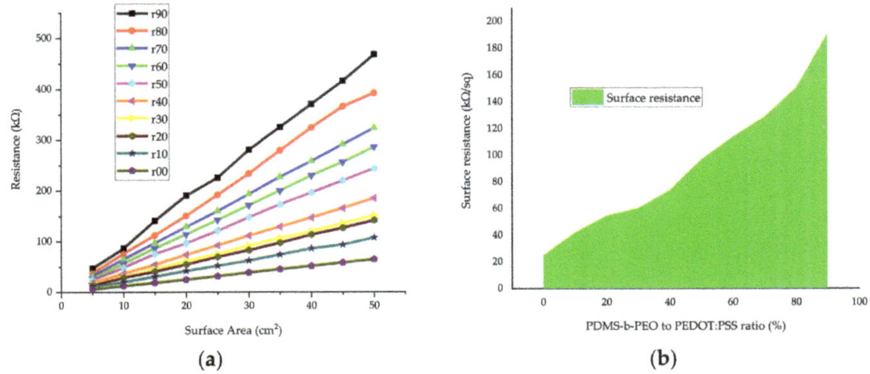

Figure 10. Effect of PDMS-b-PEO concentration on resistance: (**a**) effect of PEDOT:PSS to PDMS-b-PEO ratio on resistance at different surface area, width kept constant; (**b**) effect of PDMS-b-PEO on surface resistance.

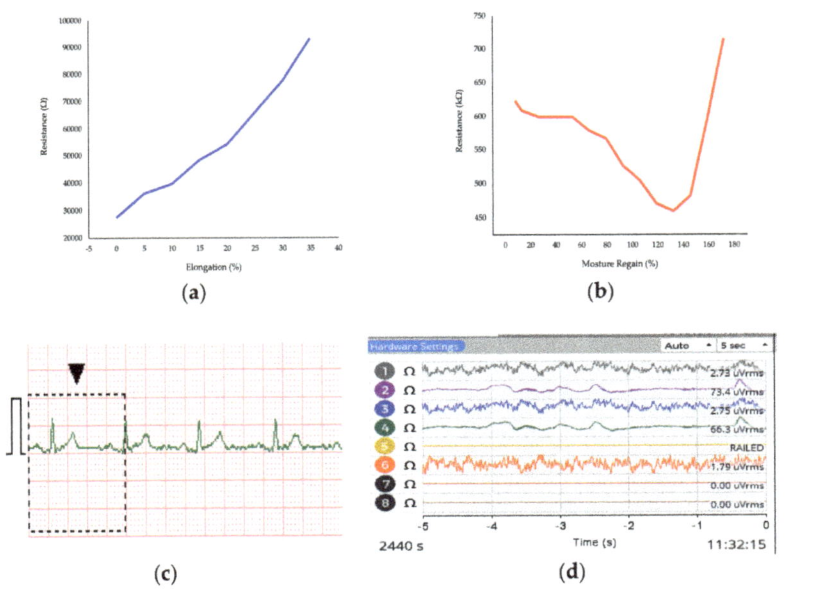

Figure 11. PEDOT:PSS/PDMS-b-PEO coated cotton electrodes for: (**a**) strain dynamic response at 0.67 cm/s rate of stretching for six seconds; (**b**) moisture dynamic response at1 ml/s rate of water spray for six seconds; (**c**) electrocardiography (ECG) signal using PC-80B; (**d**) electroencephalography (EEG) signal using OpenBCI board.

The surface resistance increased from 24.8 kΩ/sq to 96.7 kΩ/sq as the PDMS-b-PEO to PEDOT:PSS ratio increases from 0 to 90% (Figure 10b). Thus, the ratio determines the application area, as each requires specific surface resistance value. When high resistance is required, one can select a higher concentration of PEDOT-b-PEO and when lower resistance is required a lower ratio. Instead of the high resistance values obtained in this paper, one can also obtain lower resistance as mentioned in the case of an add-on of 0.013 g/cm^2, where we obtained 300 Ω/sq on the same fabric.

3.2.2. Effect of Stretching on Sensing Stability

Figure 12a shows that the stretching of the PEDOT:PSS/PDMS-b-PEO has a complex effect on the surface resistance in all samples for up to 35% elongation. In all cases, the resistance increased over the 0% elongation. This could be due to a decrease in the density of conductive components during stretch. It was observed that the change in resistance due to stretching increases at the beginning of stretching and then decreases with increasing amount of PDMS-b-PEO, with the inflection point depending on the ratio r. Whereas, when r is zero (sample r00), the percentage change in resistance increases with stretching. Moreover, the change in resistance is smaller when PDMS-b-PEO was employed. For instance, the change in resistance of sampler00 was 0.74% which is smaller than sample r60 i.e., 16% when the change in elongation was 5%. But, when the change in elongation reached 35%, the change in resistance of r60 i.e., 4% was smaller than that of r00 i.e., 131%. In addition, increasing the ratio of PDMS-b-PEO to PEDOT:PSS showed a better surface resistance recovery after stretching and better stability when re-stretched up to eight cycles. The surface resistance of sample r00 reached an infinite value at three cycles. Whereas, the most stable sample, sample r90, stayed conductive up to eight stretching cycles to its infliction point. The surface resistance of the samples after recovering from each stretch and 5 s rest is shown in Figure 12b. Though the sensing stability is not bad, it needs further improvement. A specific sample with PDMS-b-PEO can only be used to sense stretch up to the infliction point of that sample, as afterward the strain to resistivity change is unstable. Further refinement in the way of making the composite or coating technique may solve this problem.

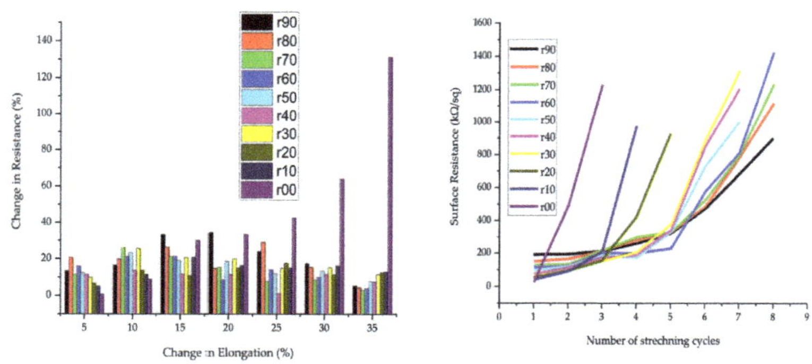

(a) (b)

Figure 12. Effect of stretching on surface resistance and sensing stability: (**a**) Change in resistance due to stretching; (**b**) surface resistance at different stretching cycles.

3.2.3. Effect of Washing on Surface Resistance

Figure 13 shows that the resistance increases after a single wash. But, the increase in surface resistance due to washing decreased from ~470% to ~30% with increasing r ratio from 0 to 90. Thus, the washing fastness improved with the increase in the amount of PDMS-b-PEO. This property of the fabric with CPC should be improved further if it is going to be used for frequently washed products.

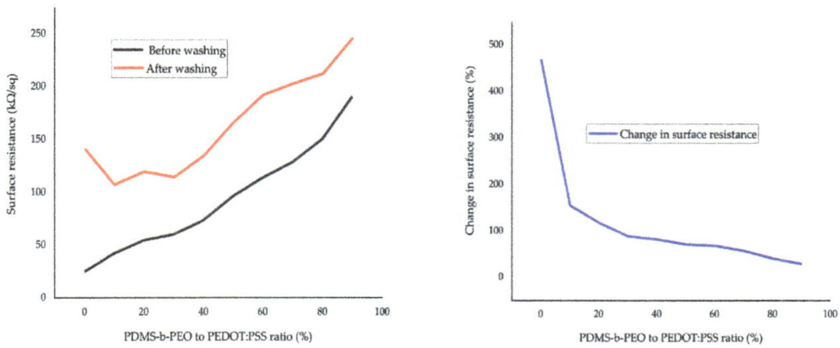

(a) **(b)**

Figure 13. Effect of washing on the surface resistance at different PDMS-b-PEO to PEDOT:PSS ratios: (**a**) surface resistance before and after washing; (**b**) percentage change in surface resistance due to washing.

4. Conclusions

In this work we have successfully developed a stretchy and flex conductive textile fabric with improved sensing stability of stretch and better washing fastness via screen printing of PEDOT:PSS and PDMS-b-PEO polymers. It was observed that the increase in the proportion of PDMS-b-PEO increases the tensile strength at break. The bending length analysis proved that the flexural rigidity drops at higher PDMS-b-PEO to PEDOT:PSS ratio which shows as the PDMS-b-PEO imparts flexibility over the pure PEDOT:PSS sample. SEM results showed that the presence of the conductive polymer composite brought smoothness and better coverage of yarn loops interstices. Samples with PEDOT:PSS/PDMS-b-PEO conductive polymer composite showed less protruding yarn loops than PEDOT:PSS alone. We found different electrical characteristics that range from 24.8 to 190.8 kΩ/sq before washing when the ratio of PDMS-b-PEO to PEDOT:PSS varies from 0 to 90%. The conductive character remained after washing, but was decreased, though much less so if PDMS-b-PEO is present. Moreover, the presence of PDMS-b-PEO improved the surface resistance recovery after release from stretching and remained conductive for more numbers of stretching cycles than sample r00.We have also realized that the surface resistance can be dropped further by increasing the add-on of the conductive polymer composite. These different electrical resistance characteristics could be chosen based on the application required to use them for sensors, interconnections, antenna, storages, and others. For instance, the conductive fabrics with higher resistance can be used to develop a textile-based humidity sensor, the middle ones can be used for the strain sensor and the lower ones can be for dry ECG, EMG, and EEG electrodes. A deep study on the electrical and physical characteristics of the developed samples is necessary to exactly indicate the correlation between the produced samples and end-use.

Author Contributions: G.B.T. conceived the idea and designed the methodology; G.B.T. wrote the paper; B.M. edited the paper and provided guidance with the experiments; G.B.T., B.M., and A.B.N. analyzed the result; K.A.F. and L.V.L. supervised and administrated the project. All authors have read and agreed to the published version of the manuscript.

Funding: The research and APC were funded by NASCERE project and Erasmus+ project Skills4Smartex.

Acknowledgments: The authors would like to express appreciation for the support of the NASCERE and Erasmus+ project Skills4Smartex. The NASCERE project was sponsored with funds from the Ethiopian government and the Erasmus+ project Skills4Smartex has been funded with support from the European Commission. The European Commission support for the production of this publication does not constitute an endorsement of the contents which reflects the views only of the authors, and the Commission cannot be held responsible for any use which may be made of the information contained therein.

Conflicts of Interest: The authors declare no conflict of interest.

References

1. Borges, B.G.A.; Holakoei, S.; das Neves, M.F.; de Menezes, L.C.; de Matos, C.F.; Zarbin, A.J.; Roman, L.S.; Rocco, M.L.M. Molecular orientation and femtosecond charge transfer dynamics in transparent and conductive electrodes based on graphene oxide and PEDOT:PSS composites. *Phys. Chem. Chem. Phys.* **2019**, *21*, 736–743. [CrossRef]
2. Güney, H.Y.; Avdan, Z.; Yetkin, H. Optimization of annealing temperature and the annealing effect on life time and stability of P3HT:PCBM-based organic solar cells. *Mater. Res. Express* **2019**, *6*, 045103. [CrossRef]
3. Hebbar, V.; Bhajantri, R.F.; Ravikumar, H.B.; Ningaraju, S. Role of free volumes in conducting properties of GO and rGO filled PVA PEDOT:PSS composite free standing films: A positron annihilation lifetime study. *J. Phys. Chem. Solids* **2018**, *126*, 242–256. [CrossRef]
4. Ding, Y.; Xu, W.; Wang, W.; Fong, H.; Zhu, Z. Scalable and Facile Preparation of Highly Stretchable Electrospun PEDOT:PSS@PU Fibrous Nonwovens toward Wearable Conductive Textile Applications. *ACS Appl. Mater. Interfaces* **2017**, *9*, 30014–30023. [CrossRef] [PubMed]
5. Sedighi, A.; Montazer, M.; Mazinani, S. Fabrication of electrically conductive superparamagnetic fabric with microwave attenuation, antibacterial properties and UV protection using PEDOT/magnetite nanoparticles. *Mater. Des.* **2018**, *160*, 34–47. [CrossRef]
6. Giuri, A.; Colella, S.; Listorti, A.; Rizzo, A.; Mele, C.; Corcione, C.E. GO/glucose/PEDOT:PSS ternary nanocomposites for flexible supercapacitors. *Compos. Part B* **2018**, *148*, 149–155. [CrossRef]
7. Hilal, M.; Han, J.I. Interface engineering of G-PEDOT:PSS hole transport layer via interlayer chemical functionalization for enhanced efficiency of large-area hybrid solar cells and their charge transport investigation. *Sol. Energy* **2018**, *174*, 743–756. [CrossRef]
8. Zajdel, T.J.; Baruch, M.; Méhes, G.; Stavrinidou, E.; Berggren, M.; Maharbiz, M.M.; Simon, D.T.; Ajo-Franklin, C.M. OPEN PEDOT:PSS-based Multilayer Bacterial-Composite Films for Bioelectronics. *Sci. Rep.* **2018**, *8*, 1–12. [CrossRef]
9. Taroni, P.J.; Santagiuliana, G.; Wan, K.; Calado, P.; Qiu, M.; Zhang, H.; Pugno, N.M.; Palma, M.; Stingelin-Stutzman, N.; Heeney, M.; Fenwick, O.; Baxendale, M.; Bilotti, E. Toward Stretchable Self-Powered Sensors Based on the Thermoelectric Response of PEDOT:PSS/Polyurethane Blends. *Adv. Funct. Mater.* **2018**, *28*, 1704285. [CrossRef]
10. Rafique, S.; Roslan, N.A.; Abdullah, S.M.; Li, L.; Supangat, A.; Jilani, A.; Iwamoto, M. UV- ozone treated graphene oxide/PEDOT:PSS bilayer as a novel hole transport layer in highly efficient and stable organic solar cells. *Org. Electrons.* **2019**, *66*, 32–42. [CrossRef]
11. Zahid, M.; Papadopoulou, E.L.; Athanassiou, A.; Bayer, I.S. Strain-responsive mercerized conductive cotton fabrics based on PEDOT:PSS/graphene. *Mater. Des.* **2017**, *135*, 213–222. [CrossRef]
12. Houghton, T.; Vanjaria, J.; Yu, H. Conductive and Stretchable Silver-Polymer Blend for Electronic Applications. In Proceedings of the 2016 IEEE 66th Electronic Components and Technology Conference (ECTC), Las Vegas, NV, USA, 31 May–3 June 2016; pp. 812–816.
13. Chen, M.; Duan, S.; Zhang, L.; Wang, Z.; Li, C. Conductive polymer composites based on graphene networks grown by chemical vapour deposition and PEDOT:PSS coating. *Chem. Commun.* **2015**, *51*, 3169–3172. [CrossRef] [PubMed]
14. Meng, Q.; Cai, K.; Du, Y.; Chen, L. Preparation and thermoelectric properties of SWCNT/PEDOT:PSS coated tellurium nanorod composite films. *J. Alloys. Compd.* **2019**, *778*, 163–169. [CrossRef]
15. Lei, K.F.; Lee, K.F.; Lee, M.Y. Development of a flexible PDMS capacitive pressure sensor for plantar pressure measurment. *Microelectron. Eng.* **2012**, *99*, 1–5. [CrossRef]
16. Lee, C.; Jug, L.; Meng, E. High strain iocompatible polydimethylsiloxane-based conductive graphene and multiwalled carbon nanotube nanocomposite strain sensors. *Appl. Phys. Lett.* **2013**, *102*, 183511. [CrossRef]
17. Kong, J.; Tong, Y.; Sun, J.; Wei, Y.; Thitsartarn, W.; Jayven, C.C.Y.; Muiruri, J.K.; Wong, S.Y.; He, C. Electrically conductive PDMS-grafted CNTs-reinforced silicone elastomer. *Compos. Sci. Technol.* **2018**, *159*, 208–215. [CrossRef]
18. Kong, K.T.S.; Mariatti, M.; Rashid, A.A.; Busfield, J.J.C. Composites: Part B Enhanced conductivity behavior of polydimethylsiloxane (PDMS) hybrid composites containing exfoliated graphite nanoplatelets and carbon nanotubes. *Compos. Part B* **2014**, *58*, 457–462. [CrossRef]

19. Gao, X.; Huang, Y.; Liu, Y.; Kormakov, S.; Zheng, X.; Wu, D.; Wu, D. Improved electrical conductivity of PDMS/SCF composite sheets with bolting cloth prepared by a spatial con fi ning forced network assembly. *RSC Adv.* **2017**, *7*, 14761–14768. [CrossRef]
20. Kim, J.; Park, J.; Jeong, U.; Park, J.W. Silver nanowire network embedded in polydimethylsiloxane as stretchable, transparent, and conductive substrates. *J. Appl. Polym. Sci.* **2016**, *133*, 43830. [CrossRef]
21. Xu, Q.; Kobayashi, S.; Ye, X.; Meng, X. Ag-PDMS composites. *Sci. Rep.* **2014**, *4*, 1–7.
22. Bian, J.; Zhou, L.; Wan, X.; Zhu, C.; Yang, B.; Huang, Y. Laser Transfer, Printing, and Assembly Techniques for Flexible Electronics. *Adv. Electron. Mater.* **2019**, *5*, 1800900. [CrossRef]
23. Cai, M.; Nie, S.; Du, Y.; Wang, C.; Song, J. Soft Elastomers with Programmable Stiffness as Strain-Isolating Substrates for Stretchable Electronics. *ACS Appl. Mater. Interfaces* **2019**, *11*, 14340–14346. [CrossRef] [PubMed]

sensors

MDPI

Article

Determining the Fracture Process Zone Length and Mode I Stress Intensity Factor in Concrete Structures via Mechanoluminescent Technology

Seong-Kyum Kim [1], Ho Geun Shin [2], Suman Timilsina [2,*] and Ji Sik Kim [2,*]

1 Department of Civil Engineering, Kumoh National Institute of Technology, 61 Daehak-ro, Gumi-si, Gyeongsangbuk-do 39177, Korea; skim@kumoh.ac.kr

2 School of Nano & Advanced Materials Engineering, Kyungpook National University, Kyeongbuk 37224, Korea; ghrms98@gmail.com

* Correspondence: sumantimilsina@knu.ac.kr (S.T.); jisikkim@knu.ac.kr (J.S.K.)

Received: 7 February 2020; Accepted: 23 February 2020; Published: 25 February 2020

Abstract: The mechanoluminescent (ML) technology that is being developed as a new and substitutive technology for structural health monitoring systems (SHMS) comprises stress/strain sensing micro-/nanoparticles embedded in a suitable binder, digital imaging system, and digital image processing techniques. The potential of ML technology to reveal the fracture process zone (FPZ) that is commonly found in structural materials like concrete and to calculate the stress intensity factor (SIF) of concrete, which are crucial for SHMS, has never been done before. Therefore, the potential of ML technology to measure the length of the FPZ and to calculate the SIF has been demonstrated in this work by considering a single-edge notched bend (SENB) test of the concrete structures. The image segmentation approach based on the histogram of an ML image as well the skeletonization of an ML image have been introduced in this work to facilitate the measurement of the length of ML pattern, crack, and FPZ. The results show ML technology has the potential to determine fracture toughness, to visualize FPZ and cracks, and to measure their lengths in structural material like concrete, which makes it applicable to structural health monitoring systems (SHMS) to characterize the structural integrity of structures.

Keywords: fracture process zone; stress intensity factor; mechanoluminescent (ML) technology; image segmentation; skeletonization; structural health monitoring

1. Introduction

The phenomenon of light emissions from certain types of materials in response to different types of mechanical forces or deformation is widely known as mechanoluminescence (ML) [1,2]. A wide range of mechanical stimuli such as tension [3], friction [4], fracture [5], compression [6], torsion [7,8], bending [9], vibration [10], impact [11], and shear [12] can induce ML from materials. Depending on the origin of the ML from a material, whether from destruction or simply from deformation, it can be broadly categorized as fractoluminescence, deformation luminescence, and triboluminescence [1]. The generation of fractoluminescence occurs during the breaking of atomic and chemical bonds in a material, which is mostly accompanied by plasma discharge; therefore, it is destructive and nonreproducible in nature. On the other hand, deformation luminescence occurs without fracturing or damaging the material; therefore, it is reproducible over a number of repetitive mechanical actions. Furthermore, deformation luminescence can be subcategorized into elasticoluminescence and plasticoluminescence depending on the emission of photons in elastic deformation and plastic deformation of the material, respectively. Since the emission of photons in elasticoluminescence has a linear relationship with applied stress or strain, it is commonly applied in stress- or strain-sensing devices. Triboluminescence, which is the third category of ML, is produced either by the contact or separation of two specific types of materials, but it has little potential for use in stress- or strain-sensing devices owing to the fact that there is no linear relationship between the applied stress amplitude and the emitted photons.

Materials that possess elasticoluminescence are widely used in the production of ML paint and ML sheets for detecting cracks and assessing the crack-tip stress field at the crack-tip vicinity lying either on the surface or on the body of a structure [12–16]. Detecting cracks, tracing the propagating path, and assessing the crack-tip stress field are important steps to be undertaken in preventing the catastrophic failure of a structure, because every material is susceptible to fracture due to repetitive and extreme mechanical forces and to the chemical and thermal environments a material faces during its lifetime. To prevent the catastrophic failure of a structure, several experimental methods have been proposed, and these are either one-dimensional or two-dimensional measurement techniques. The latter type, which includes electronic speckle interferometry [17], stereography [17], radiography [18], thermography [19], and guided wave imaging [20], is considered important as it enables direct visualization of cracks; however, all of them require both highly sophisticated measurement devices and postanalysis processes that are time-consuming and complicated. ML technology, on the other hand, comprises stress-/strain-sensing ML paint or ML sheets accompanied by an imaging system and digital image processing technique with the potential to reveal surface cracks, body cracks, and fatigue-induced microcracks [21] in addition to an ability to elucidate fracture-induced phenomena such as elastic crack-tip stress fields [12–14] and plastic crack-tip stress fields [12], and to measure stress [22] in both simple and complex structures. These advantages are why ML technology has drawn significant attention for its simplicity and cost-effectiveness as a whole-field and noncontact measurement system. Since the discovery of elasticoluminescence, ML technology has been used over the past two decades to study the fracture mechanisms in a wide range of ceramic materials such as Al_2O_3, Si_3N_4, and zirconia-based ceramics, in metallic materials such as aluminum, and in concrete [1,2,23]. Furthermore, ML has been used in the detection of multiple cracks in mega structures such as bridges [24], buildings, and pipes [25], which ensures the utility of this technology in structural health monitoring systems [26–28]. Moreover, ML paint has become an integral part of ML technology because of weatherproof characteristics that promote excellent mechanical and optical durability, which has further established its applicability in megastructures where the aforementioned features are deemed very essential.

As previously mentioned, ML technology has been applied to a wide range of structural materials to explain the crack-initiating mechanisms, crack-driving mechanisms, and the crack-tip stress fields both in elastic and plastic regions, as well as in the bridging field; however, the potential use of ML technology to visualize and explain the fracture process zones (FPZs) that are commonly found in concrete structures has never been done. In fact, the existence of FPZs in concrete structures has prevented the ability of linear elastic fracture mechanics (LEFM) to accurately address the fracture mechanisms of concrete. As a consequence, several theories concerning the effect of FPZs have been proposed [29] to elucidate the fracture mechanics of concrete. To be effective, however, these theories must be verified and upgraded, which require an accurate FPZ length [30–33]. ML technology could be used to simply fetch the FPZ dimensions. Therefore, in this work, to understand the development and advancement of FPZs via ML technology, two different types of concrete samples were considered with and without the incorporation of aggregate. To identify an FPZ and measure its length and image segmentation based on an image histogram has been introduced. Furthermore, image skeletonization was considered in order to improve the technical proficiency of measuring an FPZ and the crack length from an ML image. Finally, the crack length was measured using both a skeletonized ML image after the subtraction of FPZ and crack mouth opening displacement (CMOD: which is an indirect method to measure crack length) in order to determine the Mode I stress intensity factor (SIF) for the samples considered. The Mode I SIF is the most often used engineering design parameter in fracture mechanics which reveal the fracture tolerance of materials used in bridges, buildings, aircraft, and so forth. Both the FPZ and the SIF are important parameters that are needed to diagnose the integrity of structural materials.

2. Materials and Methods

In order to visualize and measure the FPZ length, two samples of concrete were formed with different levels of fracture toughness. Concrete from mortar only (MC) and concrete that included aggregate (AC) each were formed following normal laboratory procedure. To form the AC, cement, sand, and aggregate (diameter ~10 mm) were homogeneously combined followed by the addition of water with constant mixing for five more minutes to form a paste that was then poured into a 400 mm × 50 mm × 50 mm (Length × Breadth × Thickness) beam mold. To create a notch on the specimen, a 50 mm × 50 mm × 1 mm (Length × Breadth × Thickness) steel plate was inserted on the casting side. The MC was prepared following the same procedure without the addition of aggregate. Thereafter, the samples were allowed to stand under lab conditions (25 °C and 50% relative humidity: RH) for 12 h to allow solidification. Subsequently, the solidified samples with the dimensions shown in Figure 1a were demolded and further cured for 24 h in a chamber that maintained high levels of temperature and humidity (100 °C and 100% RH) to enhance the strength. The details regarding the concrete components are highlighted in Table 1.

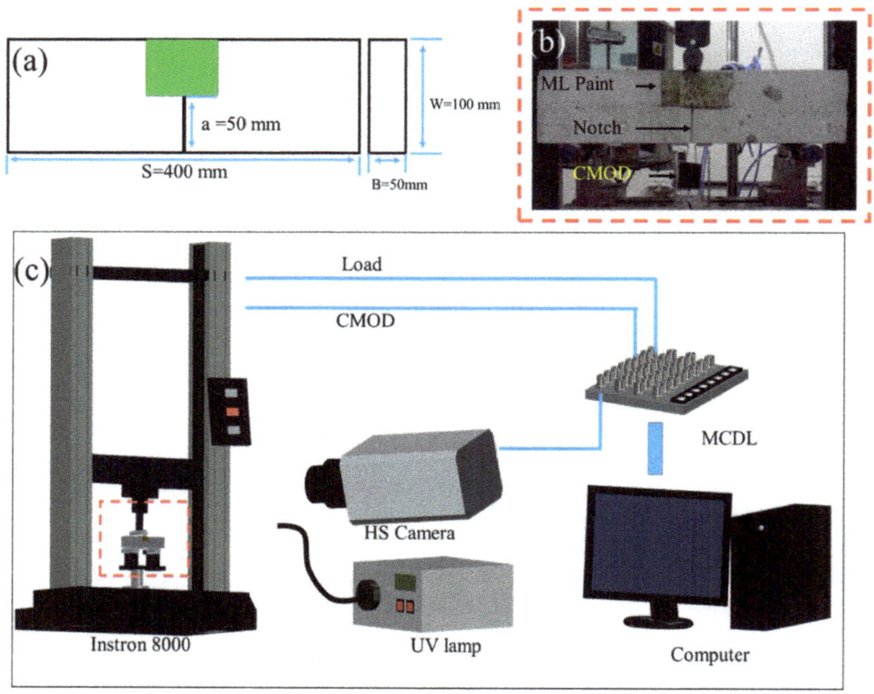

Figure 1. (**a**) Schematic diagram of a single-edge notched bend (SENB) specimen showing its dimensions with ML paint (green patch). (**b**) Real image of MC with ML paint applied to the surface ahead of the notch and a crack mouth opening displacement (CMOD) gauge attached to its notch. (**c**) Schematic diagram of the experimental setup illustrating the Instron 8000, UV lamp, high-speed camera, multichannel data link (MCDL), and computer that were used in the experiment.

Table 1. The proportion of materials used in the specimens.

Specimens	Water (kg/m^3)	Cement (kg/m^3)	Sand (kg/m^3)	Aggregate (kg/m^3)
AC	195.23	356.63	839.68	859.21
MC	195.23	356.63	1069.89	-

Stress-sensing elasticoluminescent $SrAl_2O_4$: Eu, Dy (SAOED) microparticles with an average diameter of 20 μm were purchased from NEMOTO & CO., Japan. ML paint was prepared by homogeneously mixing SAOED and optically transparent epoxy resin in a weight ratio of 3:7 at 25 °C in a centrifugal planetary mixer. The homogeneously mixed epoxy-SAOED composite was degassed for 10 min before it was applied to the samples. After applying the paint in front of the notch on the concrete sample, as illustrated in Figure 1b, the sample was allowed to stand for 24 h to allow the paint to solidify. It should be noted that ML paint thickness should be very small as compared to the thickness of concrete; otherwise, the greater thickness of paint could affect the deforming characteristics of concrete. Furthermore, the paint must be uniformly applied to avoid the effect of paint thickness on the emission. Thicker paint emits brighter ML than thinner paint; therefore, uniformity in the paint layer is important to minimize thickness effects while analyzing structural deformation.

For the SENB test, the sample was laid on a loading stage that was specially designed for SENB testing, and a CMOD gauge was attached at its notch, as shown in Figure 1b. The paint on the sample was exposed to UV light for 5 min and then aged in the dark for 2 min to allow the long phosphorescence to relax to a reasonable level. After this, a 2000 N/min load was applied until the concrete fractured. During loading, the painted surface was photographed on a macro scale at a frame speed of 250 frames/s using a high-speed imaging system. The data recorded from a load cell and CMOD were synchronized with each frame using a multichannel data link (MCDL). The experimental setup is schematically illustrated in Figure 1c.

3. Results and Discussion

The load-CMOD response curves of AC and MC illustrating their levels of fracture strength under SENB testing are illustrated in Figure 2. The load first increased rapidly to the peak level and then dropped with the propagation of cracks in both samples. It is well understood that cracks in concrete tend to move along a path either where the interfacial zone is weakest or through any existing pores. If a crack encounters aggregate particles, it is forced either to move through the stronger aggregate or to deflect and propagate around the weaker aggregate–mortar interface. In both situations, a large amount of energy is absorbed; consequently, the AC showed fracture strength that was higher than that of the MC.

Figure 3 shows representative sequential ML images of crack initiation and propagation in two samples. The images at 0 N showed faint emissions, which was attributed to the long phosphorescence of SAOED. The advancing ML pattern in AC and MC was vivid at peak loads of 1.46 and 1.04 kN, respectively, as well as for onward loads. Even before the peak load, a faint ML pattern that developed at the notch tip was apparent in both specimens at load levels of 0.2 and 0.1 kN, respectively, in both AC and MC. The ML pattern in the AC was much higher than that in the MC, which resulted from the difference in fracture strength. The length of the ML pattern was then compared with the crack length obtained from the CMOD gauge, as illustrated in Figure 4. Figure 4a compares the ML pattern with the crack length calculated from the CMOD for the AC while Figure 4b shows that of the MC. In both samples, the overall ML pattern length was higher than the crack length that was calculated based on the CMOD gauge for a given load. This difference clearly shows that the ML pattern was not generated solely from the crack but also from the microcracking region, also known as the FPZ, ahead of the traction-free crack tip, which the CMOD does not measure.

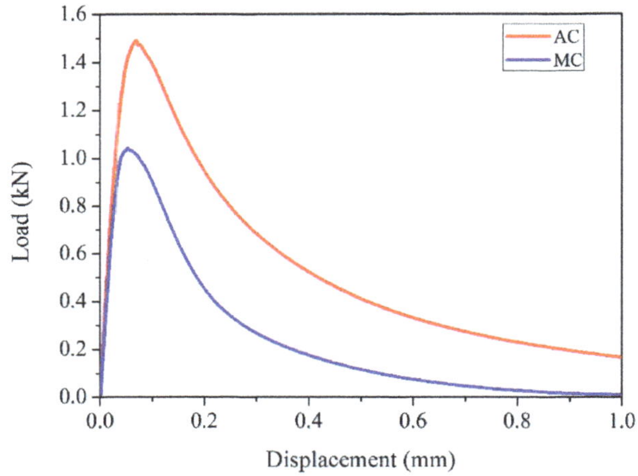

Figure 2. Load-CMOD relation curves of AC and MC illustrating their different levels of fracture strength.

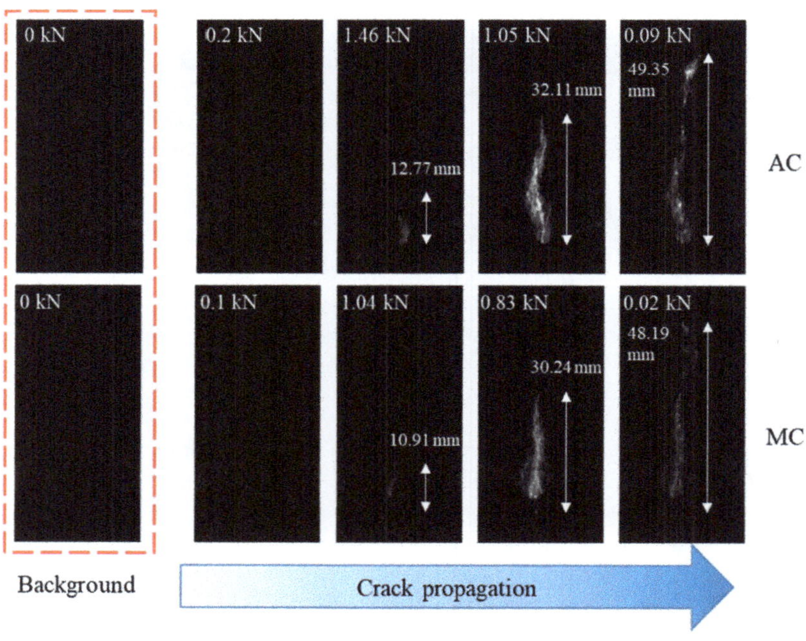

Figure 3. Sequential ML gray images illustrating crack initiation and propagation in AC and MC during SENB testing, which were recorded using a high-speed camera at a frame speed of 250 frames/s.

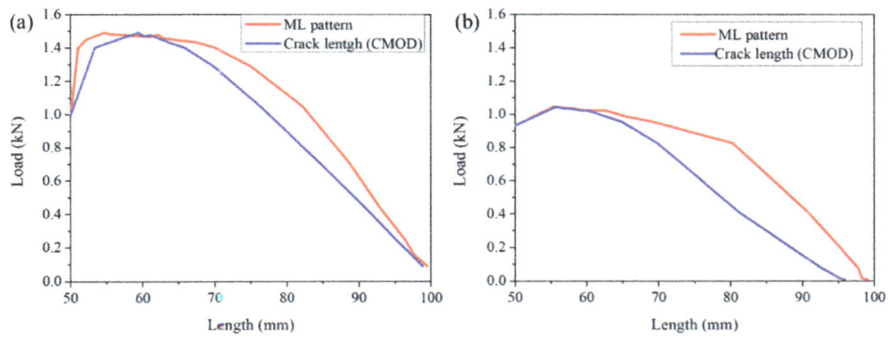

Figure 4. Comparison of the ML pattern length and crack length calculated using the indirect method (CMOD) for AC (**a**) and MC (**b**).

The CMOD was measured using TML Dynamic Strainmeters with sampling frequency of 5 kHz. The CMOD data from TML Dynamic Strainmeters was further synchronized with camera images and applied load using MCDL with 250 sample/s. Therefore, each image was taken for every 0.004 s and the total duration of whole experiment shown in Figure 2 was 2.844 s. Since the concrete is brittle in nature, the crack speed under 2000 N/min loading rate was very high resulting in very short duration for crack propagation (around 0.064 s). As a result, out of 500 synchronized data points, around 16 images only depicted crack propagation, which were considered for the assessment of the FPZ.

Understanding the FPZ is essential due to the invalid nature of linear elastic fracture mechanics (LEFM), inaccuracies in the predictions of failure for concrete structures, and the fracture energy dependency on FPZ size. Therefore, an accurate size for FPZ is helpful when using the finite element method (FEM) to simulate the crack propagation path and for the development of new theoretical models [29–33]. Techniques that involve X-ray, acoustic, interferometry, and optical microscopic analyses have been employed to elucidate the FPZ phenomenon. The aforementioned methods, however, are time-consuming and inaccurate, and they require testing instruments that generate mechanical vibration, require advanced sample preparation techniques, and are prohibitively expensive. To overcome these disadvantages, digital image correlation (DIC) techniques have been employed by some researchers where digital images obtained from high-speed imaging systems are analyzed to measure the FPZ based on strain or displacement fields [33]. Both DIC and ML technologies use an imaging system to record the fracture process and the postprocessing of digital images to determine various fracture parameters, which further promotes the use of ML images to assess the FPZ.

Before the estimation of FPZ length, it is first necessary to understand the stress-field distribution around the crack-tip vicinity in concrete. The stress profile in the front of the crack tip along the crack-propagating axis in concrete is shown schematically in Figure 5a, where the peak stress lies in the FPZ rather than at the crack tip. A shift in the peak stress away from the crack tip has been observed in elastic–plastic material via ML technology. The length of the shift in the peak stress determined using ML technology approximated the length obtained using Irwin's method for finding a virtual crack tip based on the stress intensity factor of the material [12]. Unlike isotropic elastic–plastic material where the stress field is concentrated at a particular point, the inhomogeneity characteristics of concrete and the distributions of several microcracks, as shown in schematic Figure 5a, show that the stress field is highly distributed around microcracks, which makes it difficult to pinpoint the peak stress point at the FPZ when using ML technology. However, the FPZ stress field should trigger an ML pattern, as the lighter green color reveals in Figure 5b. In fact, the stress field in an FPZ is transferred to stress-sensing, microsized SAOED particles through the binder, which then triggers an emission from the SAOED particles that results in the formation of an ML pattern. The photon emission from SAOED can be described using a trap-controlled mechanism [1]. An electron from Eu^{2+} is released to conduction band during UV irradiation and gets trapped by oxygen vacancy. Thereafter, the trapped electron is released by mechanical stimuli and is recombined with the oxidized Eu^{3+}, which results in the emission of photons, consequently creating ML patterns [1,2,23]. The larger length of ML pattern than the length of crack measured using CMOD clearly indicates an emission from the microcracking zone as clarified in Figure 5; however, it is difficult to discern the difference between an emission from the propagated crack and one from the FPZ. In principle, the emission that originates from the crack propagation should be higher than the emission from a microcracking zone since stress concentration after the coalescence of microcracks will be localized and consequently results in a higher level of emission. However, due to dispersion of the stress fields in the FPZ, emissions from the FPZ should be fainter compared with those from the main crack, as illustrated in Figure 5b. To distinguish between an emission from a main crack and that from a microcracking zone, it is necessary to adopt a digital image analysis technique; therefore, image segmentation based on an image histogram was adopted to distinguish the emission patterns in this work.

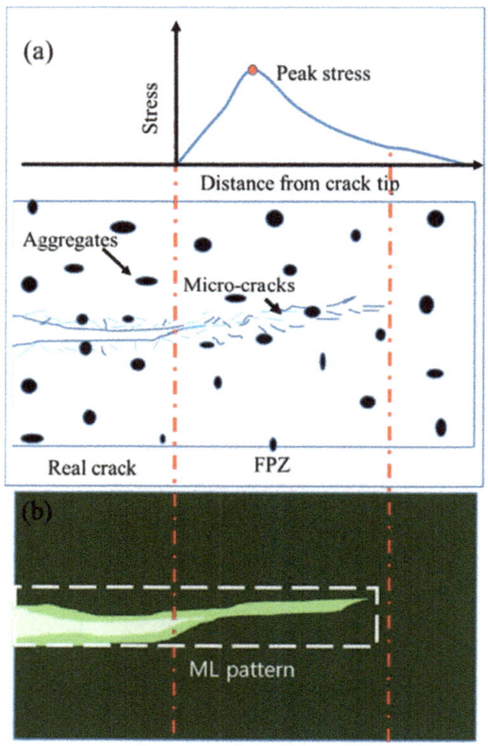

Figure 5. (**a**) Stress profile ahead of a crack tip in concrete material where the FPZ has the maximum stress concentration. (**b**) Schematic diagram showing the ML intensity from ML paint on a concrete sample. Among the three different colors, the dark green illustrates long phosphorescence, the lightest green depicts the emission from a recently propagated crack, and the remaining lighter green is the emission from a microcracking zone.

An image histogram is a graph of the pixel intensity (on the x-axis) versus the number of pixels (on the y-axis) [34,35]. The histogram of an ML image (Figure 6a) is depicted in Figure 6b, where the x-axis represents the gray level depicting the intensity value and the y-axis represents the total number of pixels. From a histogram, an ML image can be broadly categorized into three different parts: the First Zone (0–51), the Second Zone (52–178), and the Third Zone (179–256). The variation in pixel intensity is also depicted by the small square-shaped image parts in Figure 6b. The First Zone is free from any stress and any emission from this zone originates from the enduring phosphorescence of the ML paint. The Second Zone has an intensity gradient that is distinctive from those of both the First and Third Zones, and is believed to originate from the microcracking zone around the main crack. A wide range of pixel values in this zone suggest a nonuniform stress distribution owing to the fact that there typically are different stress fields in the vicinity of a microcrack and between the microcracks in a region. Moreover, the densities of microcracks vary as they extend away from the traction-free main crack, which also results in a degrading of the intensity away from the traction-free main crack. The Third Zone shows the emission from the main crack, where the stress localization should result in the brightest zone in the ML image. Based on this histogram, the image is segmented into three respective zones as illustrated by the different colors in Figure 6c [36,37]. The First Zone is highlighted in green, the Second Zone is highlighted in blue, and the Third Zone is highlighted in red. The Second and Third Zones are also shown separately in Figure 6d,e, respectively. In the segmented Figure 6d,e, the length of the FPZ can be calculated by measuring the gap length.

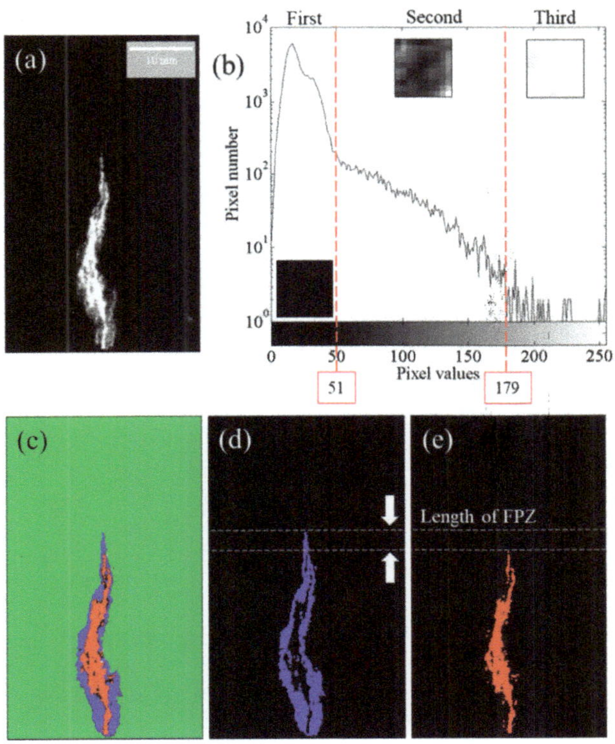

Figure 6. The histogram of (**a**) has been derived using MATLAB software, which is illustrated in (**b**). Based on the histogram, (**a**) has been segmented into three different zones as illustrated in (c) by green, blue, and red colors. The Second Zone and the Third Zone are illustrated separately in (**d**,**e**), respectively; meanwhile, the length of the FPZ is illustrated from (**d**,**e**).

In order to measure the crack length based on the ML image, it is important to develop new methods to reduce time and increase accuracy, which is impossible with a manual measurement using the ruler in Photoshop. For example, in an isotropic composite material where a clear localized crack-tip stress field distribution is observable from the ML, the crack tip is easily determined because the maximum ML emission lies at the crack tip, which is the site of the maximum concentration of stress. Therefore, knowing the crack-tip position simplifies the measurement of the crack length and allows the instantaneous calculation of fracture parameters such as the J integral and the SIF when using MATLAB software [12]. The nonhomogeneous nature of a concrete surface, however, can result in the aforementioned nonconcentrated and spatially distributed stress fields, which can make it impossible to determine a crack tip akin to the experience of working with isotropic homogeneous materials. Therefore, a new method is essential. In the present work, crack and FPZ measurements using skeleton images of the ML were introduced in order to facilitate their automated measurement.

Skeletonization provides an effective and compact representation of an object by reducing its dimensionality to a "medial axis" or "skeleton" while preserving the topologic and geometric properties of the object. Since Blum introduced the concept of skeletonization [38], it has been applied to numerous forms of image processing and computer-vision applications such as object description, matching, classification, and so forth. For skeletonization, several different types of computational approaches exist, and their fundamental principles are mentioned in the literature [39,40]. In fact, many researchers have undertaken skeletonization based on continuous approaches, while others have preferred purely digital methods for skeletonization. In this work, skeletonization was undertaken using the MATLAB library code, and the results are illustrated in Figure 7.

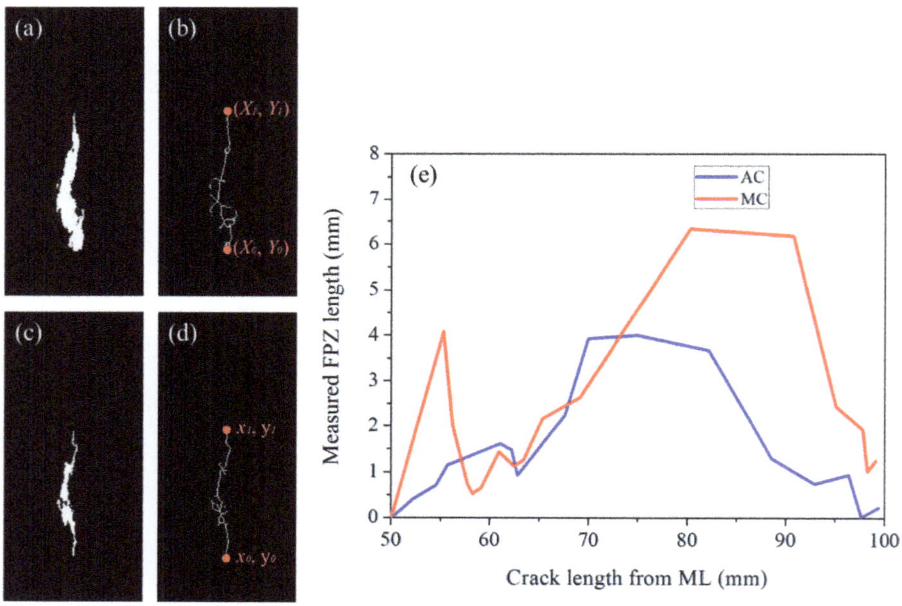

Figure 7. (**a**) Binary ML image of Figure 6a showing the ML pattern. (**b**) Skeleton image of (**a**). (**c**) Binary image of Figure 6e showing the crack after the exclusion of the First Zone and the Second Zone. (**d**) Skeleton image of (**c**). (**e**) Comparison of the FPZs of two different samples with crack lengths obtained from the skeleton images.

The first step involves changing ML images from a high-speed imaging system into binary images, after which the skeleton images are obtained using the MATLAB library code. A representative binary image and its skeleton image are depicted in Figure 7a,b, respectively. Then, segmented images showing the crack are converted into binary images and subsequently into their skeleton images. A representative binary image and its skeleton image are depicted in Figure 7c,d, respectively. The length of the ML pattern in Figure 7a, which consists of both an actual crack and its FPZ, is measured simply from the ends of the skeleton coordinates along the *y*-axis, Y_1-Y_o.The crack length in Figure 7c was calculated in a similar fashion, y_1-y_o. Thereafter, the length of the FPZ was determined by subtracting y_1 from Y_1 (i.e., Y_1-y_1). Therefore, measuring the crack length and FPZ using skeletonized images is much easier; moreover, the accuracy is increased since a length equal to a single pixel can be measured accurately. In fact, measuring the ML pattern with spatial distribution in the *xy* plane is a tough job for the naked eye using Photoshop's ruler, because it is not easy to determine the ends of the ML pattern, which results in errors as well as in a great requirement of time and effort. These facts are why approaches of image skeletonization have been undertaken by other researchers to measure the crack length and visualize the crack path [41]; however, the FPZ length has never been measured, particularly in the ML field.

Following the aforementioned procedure, crack and FPZ lengths were measured for both samples in the present study, as illustrated in Figure 7e. The illustration shows that with the advancement of the crack, the FPZ increased first and then dropped in both samples. The FPZ length that was measured using the DIC technique showed the same trend, but it monotonically increased and dropped without fluctuating as in this method [33]. The fluctuating FPZ length with crack advancement likely originated from the nonhomogeneous nature of concrete in conjunction with the high sensitivity of ML paint. In fact, a highly sensitive type of paint or patch is desirable for the sake of understanding the crack vicinity fracture phenomenon of a structure under a static or dynamic loading environment. For example, paint revealed the FPZ before the advancement of the crack, as illustrated in Figure 3. At this stage, the CMOD showed a crack length almost equal to the initial notch length, but paint revealed the development of an FPZ. This information is essential for structural health monitoring and in fact more advanced stress-sensing paint is desirable so that stress distribution at the crack vicinity can be visualized even without the early development of an FPZ. The realization of such paint is possible by using nano-/microsized stress-sensing particles with the ability to respond to conditions of low stress; unfortunately, one of the best stress-sensing particles used in this type of work emits photons after a threshold stress level of 5 MPa [13].

After the determination of an FPZ, it is worthwhile to determine the stress intensity factor of the samples based on the crack length using an indirect method as well as the crack length using the image after the exclusion of the FPZ. The SIF was calculated using Equation (1), which is a commonly used SENB test for concrete [42].

$$SIF = K_I = (\frac{3PS}{2W^2B})\sqrt{\pi a}.f(\frac{a}{W}) \tag{1}$$

where
P = applied load
S = span
W = specimen depth
B = specimen thickness
a = crack length

$$f\left(\frac{a}{W}\right) = \frac{1.99 - \left(\frac{a}{W}\right)\left(1-\frac{a}{W}\right)\left[2.15 - 3.93\frac{a}{W} + 2.70\left(\frac{a}{W}\right)^2\right]}{\sqrt{\pi}\left(1+2\frac{a}{W}\right)\left(1-\frac{a}{W}\right)^{\frac{3}{2}}} = \text{geometric factor} \tag{2}$$

The results are illustrated in Figure 8. Figure 8a compares SIFs obtained from the indirect method and from the ML crack. The agreement between the two methods was as good as a/W= 0.7; however, once the crack propagated further, the SIFs diverged. Unlike in AC, the SIFs in MC showed perfect agreement between the two methods. The divergence of the SIF is directly related to the slight mismatches between the crack lengths obtained from the two different methods. In fact, for a given load and crack length, the SIF increases geometrically with an increase in the a/w, as indicated by Equation (2), and, therefore, despite a slight mismatch in the crack length, the SIF tends to show a gap, and this gap increases with a higher a/W.

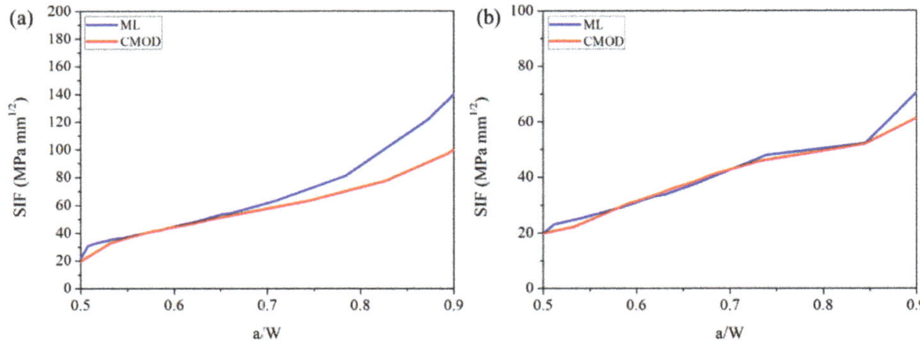

Figure 8. Comparison of SIFs that were obtained from an indirect method (CMOD) and by using ML technology, where (**a**) shows the SIFs of AC and (**b**) shows the SIFs of MC.

This work has clearly shown that ML technology can reveal the development and advancement of the FPZ that is most commonly found in concrete structures as an addendum to the actual measurement of real crack length. Also, the SIF, which is a very important material characteristic, can be calculated using ML technology. Using CMOD can only provide information about crack length, but ML technology can provide full-field information about the crack path, crack tip, multiple cracks, and the distribution of the stress field at the crack-tip vicinity in real time. These features are desired in developing a structural health monitoring system to characterize the structural integrity of large structures such as bridges and buildings [24,43]. ML technology can be a reliable and effective tool for SHMS because it is a full-field and noncontacting measurement system with a short response time that is cost-effective and offers simplicity in application and data processing. Nonetheless, there is always room for enhancing the effectiveness of ML technology which mainly depends on a few improvements: first, the discovery of noble stress/strain-sensing novel particles with very low threshold stress, so that the emission of photons is guaranteed under minimally applied stress; second, the discovery of optimized binders where the stress transferability from binder to stress-sensing particles is high; as a result, the emission of photons is enhanced remarkably. Last, the incorporation of advanced digital image processing techniques is necessary, so that several features of fracture mechanics can be revealed effectively in less time via ML image analysis.

4. Conclusions

A new method to measure the FPZ length and calculate the SIF based on ML technology has been demonstrated in this work by considering the SENB testing of two different concrete structures with and without aggregate. The length of the FPZ was calculated by subtracting the visualized crack length from the visualized ML pattern after the ML image segmentation into a main crack and microcrack sections. The image segmentation was accomplished by adopting an image histogram in this work and was a very effective method despite its simplicity. To more simply measure the crack length and the FPZ, a skeletonizing technique was introduced, which was very useful in measuring the length of the patterns in the ML images. The measured FPZ showed a tendency to first increase and then drop, which was a trend reported by other studies. Fluctuation in the increase and decrease in the FPZ length was attributed to the sensitivity of the ML paint as it revealed the inhomogeneous nature of a concrete structure. Furthermore, another important fracture parameter, SIF, was calculated based on the crack length measured using both the ML technology and the indirect CMOD method. Even though the SIFs from both methods showed a perfect match in the sample without aggregate, a slight mismatch in crack length resulted in different values of SIFs for the two different methods in the sample with aggregate. Nonetheless, as a whole-field and noncontact method to determine fracture parameters such as fracture toughness and FPZ length in structural material like concrete, ML technology has enhanced its potential use in structural health monitoring systems (SHMS) to diagnosis the structural integrity of megastructures.

Author Contributions: Conceived and designed the experiments, J.S.K.; Performed the experiments, S.-K.K. and S.T.; Analyzed the experimental data, S.-K.K. and S.T.; Investigation, H.G.S.; Visualization, H.G.S.; Writing—original draft preparation, S.-K.K. and S.T.; Writing—review & editing, S.-K.K., S.T. and J.S.K.; Funding acquisition, S.-K.K. and J.S.K. All authors have read and agreed to the published version of the manuscript.

Funding: This work was supported by a National Research Foundation of Korea (NRF) grant funded by the Korean government (MSIT) (grant numbers: 2017R1A2B4001877 and 2019R1A2C1011264). This work was also supported by a grant (14RDRP-B076268) from Regional Development Research Program funded by Ministry of Land, Infrastructure and Transport of Korean government.

Acknowledgments: The above financial support is greatly appreciated.

Conflicts of Interest: The authors declare no conflict of interest.

References

1. Zhang, J.-C.; Wang, X.; Marriott, G. Trap-controlled mechanoluminescent materials. *Prog. Matter. Sci.* **2019**, *103*, 678–742. [CrossRef]
2. Feng, A.; Smet, P.F. A review of mechanoluminescence in inorganic solids: Compounds, mechanisms, models and applications. *Materials.* **2018**, *11*, 484. [CrossRef] [PubMed]
3. Rahimi, M.R.; Yun, G.J.; Choi, J.-S. A predictive mechanoluminescence transduction model for thin-film $SrAl_2O_4$: Eu^{2+}, Dy^{3+} (SAOED) stress sensor. *Acta. Mater.* **2014**, *77*, 200–211. [CrossRef]
4. Matsui, H.; Xu, C.-N.; Liu, Y.; Tateyama, H. Origin of mechanoluminescence from Mn-activated $ZnAl_2O_4$. *Phys. Rev. B.* **2004**, *69*, 235109. [CrossRef]
5. Chandra, B.P.; Chandra, V.K.; Jha, P. Models for intrinsic and extrinsic fracto-mechanoluminescence of solids. *J. Lumin.* **2013**, *135*, 139–153. [CrossRef]
6. Xu, C.-N.; Watanabe, T.; Akiyama, M.; Zheng, X.-G. Direct view of stress distribution in solid by mechanoluminescence. *Appl. Phys. Lett.* **1999**, *74*, 2414–2416. [CrossRef]
7. Kim, G.-W.; Kim, J.-S. Dynamic torsional response analysis of mechanoluminescent paint and its application to non-contacting automotive torque transducers. *Meas. Sci. Technol.* **2013**, *25*, 015009. [CrossRef]
8. Kim, J.S.; Kim, G.-W. New non-contacting torque sensor based on the mechanoluminescence of ZnS: Cu microparticles. Sensor. *Actuat. A-Phys.* **2014**, *218*, 125–131. [CrossRef]
9. Jia, Y.; Yei, M.; Jia, W. Stress-induced mechanoluminescence in $SrAl_2O_4$: Eu^{2+}, Dy^{3+}. *Opt. Mater.* **2006**, *28*, 974–979. [CrossRef]
10. Zhan, T.; Xu, C.-N.; Fukuda, O.; Yamada, H.; Li, C. Direct visualization of ultrasonic power distribution using mechanoluminescent film. *Ultrason. Sonochem.* **2011**, *18*, 436–439. [CrossRef]
11. Zhan, T.Z.; Xu, C.N.; Yamada, H.; Terasawa, Y.; Zhang, L.; Iwase, H.; Kawai, M. Enhancement of impact-induced mechanoluminescence by swift heavy ion irradiation. *Appl. Phys. Lett.* **2012**, *100*, 014101. [CrossRef]
12. Basnet, R.; Timilsina, S.; Lee, K.H.; Kim, J.S. Evaluation of the elasto-plastic crack tip singularities via mechano-luminescent effects. *Int. J. Eng. Sci.* **2018**, *123*, 127–142. [CrossRef]
13. Timilsina, S.; Lee, K.H.; Jang, I.Y.; Kim, J.S. Mechanoluminescent determination of the mode I stress intensity factor in $SrAl_2O_4$: Eu^{2+}, Dy^{3+}. *Acta. Mater.* **2013**, *61*, 7197–7206. [CrossRef]
14. Timilsina, S.; Lee, K.H.; Kwon, Y.N.; Kim, J.S. Optical evaluation of in situ crack propagation by using mechanoluminescence of $SrAl_2O_4$: Eu^{2+}, Dy^{3+}. *J. Am. Ceram. Soc.* **2015**, *98*, 2197–2204. [CrossRef]
15. Timilsina, S.; Bashnet, R.; Kim, S.H.; Lee, K.H.; Kim, J.S. A life-time reproducible mechano-luminescent paint for the visualization of crack propagation mechanisms in concrete structures. *Int. J. Fatigue* **2017**, *101*, 75–79. [CrossRef]
16. Fujio, Y.; Xu, C.-N.; Terasawa, Y.; Sakata, Y.; Yamabe, J.; Ueno, N.; Terasaki, N.; Yoshida, A.; Watanabe, S.; Murakami, Y. Sheet sensor using $SrAl_2O_4$: Eu mechanoluminescent material for visualizing inner crack of high-pressure hydrogen vessel. *Int. J. Hydrogen. Energ.* **2016**, *41*, 1333–1340. [CrossRef]
17. Kim, K.-S.; Kang, K.-S.; Kang, Y.-J.; Cheong, S.-K. Analysis of an internal crack pressure pipeline using ESPI and stereography. *Opt. Laser. Technol.* **2003**, *35*, 639–643. [CrossRef]

18. Kazantsev, I.G.; Lemahieu, I.; Salov, G.I.; Denys, R. Statistical detection of defects in radiographic images in nondestructive testing. *Signal. Process.* **2002**, *82*, 791–801. [CrossRef]
19. Vageswar, A.; Balasubramaniam, K.; Krishnamurthy, C.V.; Jayakumar, T.; Raj, B. Periscope infrared thermography for local wall thinning in tubes. *NDT&E. Int.* **2009**, *42*, 275–282.
20. Hayashi, T.; Murase, M. Defect imaging with guided waves in a pipe. *J. Acoust. Soc. Am.* **2005**, *117*, 2134–2174. [CrossRef]
21. Terasaki, N.; Xu, C.-N. Historical-log recording system for crack opening and growth based on mechanoluminescent flexible sensor. *IEEE. Sens. J.* **2013**, *13*, 3999–4004. [CrossRef]
22. Kim, J.S.; Kwon, Y.-N.; Sohn, K.S. Dynamic visualization of crack propagation and bridging stress using the mechano-luminescence of $SrAl_2O_4$: (Eu, Dy, Nd). *Acta. Mater.* **2003**, *51*, 6437–6442. [CrossRef]
23. Timilsina, S.; Kim, J.S.; Kim, J.H.; Kim, G.-W. Review of State-of-the-Art Sensor Applications Using Mechanoluminescence Microparticles. *Int. J. Precis. Eng. Manuf.* **2016**, *2016*, 1237–1247. [CrossRef]
24. Liu, L.; Xu, C.-N.; Yoshida, A.; Tu, D.; Ueno, N.; Kainuma, S. Scalable elasticoluminescent strain sensor for precise dynamic stress imaging and onsite infrastructure diagnosis. *Adv. Mater.* **2019**, *4*, 1800336. [CrossRef]
25. Ono, D.; Xu, C.-N.; Li, C.; Bu, N. Visualization of Internal Defect of a Pipe Using Mechanoluminescent Sensor. *Jpn. Soc. Exp. Mech.* **2010**, *10*, 152–156.
26. Olawale, D.O.; Sullivan, G.; Dickens, T.; Tsalickis, S.; Okoli, O.I.; Sobanjo, J.O.; Wang, B. Development of a triboluminescence-based sensor system for concrete structures. *Struct Health Monit.* **2012**, *11*, 139–147. [CrossRef]
27. Olawale, D.O.; Uddin, J.M.; Yan, J.; Dickens, T.J.; Okoli, O.O.I. Triboluminescent Sensors for Cement-Based Composites. In *Triboluminescence*; Springer: Berlin/Heidelberg, Germany, 2016.
28. Peng, D.; Chen, B.; Wang, F. Recent Advances in Doped Mechanoluminescent Phosphors. *ChemPlusChem* **2015**, *80*, 1209–1215. [CrossRef]
29. Wittmann, F.H.; Hu, X. Fracture process zone in cementitious materials. *Int. J. Fract.* **1991**, *51*, 3–18. [CrossRef]
30. Skarzynski, L.; Tejchman, J. Calculations of fracture process zones on meso-scale in notched concrete beams subjected to three-point bending. *Eur. J. Mechan. A Solids.* **2010**, *29*, 746–760. [CrossRef]
31. Otsuka, K.; Date, H. Fracture process zone in concrete tension specimen. *Eng. Fract. Mechan.* **2000**, *65*, 111–131. [CrossRef]
32. Guo, Z.K.; Kobayashi, A.S.; Hawkins, N.M. Further studies on fracture process zone for mode I concrete fracture. *Eng. Fract. Mechan.* **1993**, *46*, 1041–1049.
33. Wu, Z.; Rong, H.; Zheng, J.; Xu, F.; Dong, W. An experimental investigation on the FPZ properties in concrete using digital image correlation technique. *Eng. Fract. Mechan.* **2011**, *78*, 2978–2990. [CrossRef]
34. Tan, L.; Jiang, J. *Digital Signal Processing*, 3rd ed.; Academic Press: Cambridge, MA, USA, 2018.
35. Bovik, A.C. *The Essential Guide to Image Processing*, 2nd ed.; Academic Press: Cambridge, MA, USA, 2009.
36. Caraiman, S.; Manta, V.I. Histogram-based segmentation of quantum images. *Theor. Comput. Sci.* **2014**, *529*, 46–60. [CrossRef]
37. Qin, K.; Xu, K.; Du, Y.; Li, D. An image segmentation approach based on histogram analysis utilizing cloud model. In Proceedings of the 2010 Seventh International Conference on Fuzzy Systems and Knowledge Discovery (FSKD 2010), Yantai, China, 10–12 August 2010; pp. 524–528.
38. Blum, H.F. A transformation for extracting new descriptors of shape. In *Models for the Perception of Speech and Visual Form*; MIT Press: Cambridge, MA, USA, 1967.
39. Siddiqi, K.; Prizer, S.M. Mathematics, Algorithms and Applications. In *Medial Representations*; Springer: Berlin/Heidelberg, Germany, 2008.
40. Jin, D.; Iyer, K.S.; Chen, C.; Hoffman, E.A.; Saha, P.K. A robust and efficient curve skeletonization algorithm for tree-like objects using minimum cost paths. *Pattern. Recog. Lett.* **2016**, *76*, 32–40. [CrossRef] [PubMed]
41. Jiang, D.; Li, G.; Sun, Y.; Kong, J.; Tao, B.; Zhou, D.; Chen, D.; Ju, Z. Visual-Based Crack Detection and Skeleton Extraction of Cement Surface. In Proceedings of the 12th International Conference on Intelligent Robotics and Applications (ICIRA 2019), Shenyang, China, 11–18 August 2019; pp. 541–552.

42. Shah, S.P.; Swartz, S.E.; Ouyang, C. *Fracture Mechanics of Concrete: Applications of Fracture Mechanics to Concrete, Rock and Other Quasi-Brittle Materials*; John Wiley & Sons Inc.: Hoboken, NJ, USA, 1995.
43. Burgos, D.A.T.; Vargas, R.C.G.; Cesar Pedraza, C.; Agis, D.; Pozo, F. Damage Identification in Structural Health Monitoring: A Brief Review from its Implementation to the Use of Data-Driven Applications. *Sensors* **2020**, *20*, 733. [CrossRef] [PubMed]

sensors

MDPI

Article

Impact of Fabric Properties on Textile Pressure Sensors Performance

Luca Possanzini [1], Marta Tessarolo [1,*], Laura Mazzocchetti [2], Enrico Gianfranco Campari [1] and Beatrice Fraboni [1]

[1] Department of Physics and Astronomy, University of Bologna, Viale Berti Pichat 6/2, 40127 Bologna, Italy; luca.possanzini2@unibo.it (L.P.); enrico.campari@unibo.it (E.G.C.); beatrice.fraboni@unibo.it (B.F.)

[2] Department of Industrial Chemistry, University of Bologna, Via Risorgimento 4, 40136 Bologna, Italy; laura.mazzocchetti@unibo.it

* Correspondence: marta.tessarolo3@unibo.it; Tel.: +39-051-2099-5315

Received: 11 September 2019; Accepted: 22 October 2019; Published: 28 October 2019

Abstract: In recent years, wearable technologies have attracted great attention in physical and chemical sensing applications. Wearable pressure sensors with high sensitivity in low pressure range (<10 kPa) allow touch detection for human-computer interaction and the development of artificial hands for handling objects. Conversely, pressure sensors that perform in a high pressure range (up to 100 kPa), can be used to monitor the foot pressure distribution, the hand stress during movements of heavy weights or to evaluate the cyclist's pressure pattern on a bicycle saddle. Recently, we developed a fully textile pressure sensor based on a conductive polymer, with simple fabrication and scalable features. In this paper, we intend to provide an extensive description on how the mechanical properties of several fabrics and different piezoresistive ink formulation may have an impact in the sensor's response during a dynamic operation mode. These results highlight the complexity of the system due to the presence of various parameters such as the fabric used, the conductive polymer solution, the operation mode and the desired pressure range. Furthermore, this work can lead to a protocol for new improvements and optimizations useful for adapting textile pressure sensors to a large variety of applications.

Keywords: textile pressure sensor; e-textile; dynamic mode; PEDOT:PSS

---

1. Introduction

In recent years, the field of wearable technology has attracted great attention and different kinds of sensors have been developed. The advent of this new technology is changing the way people interface with the external world. Wearable electronic sensors are growing very popular in several fields such as medical [1], healthcare [2–4], wellness [5], entertainment [6,7] and safety [8]. The most requested applications deal with wearable activity trackers that provide a powerful and detailed self-monitoring as well as an opportunity to directly control personal habits and behaviors. Such devices are usually based on inorganic electronics chips and are considered wearable because they are in contact with the body, typically the wrist. However, the most promising completely wearable and imperceptible sensors are those fabricated directly on fabric, that is, fully textile physical or chemical sensors. Fabrics, with their comfort and fitting features, are the best platforms and substrate onto which to directly realize a wearable sensor device and, when provided with sensing properties, are named *Smart Textiles*. They can be made sensitive to various external stimuli, both physical and chemical [9], such as the changes in force [10,11], pressure [12,13], deformation [14,15], temperature [16] or the concentration of specific compounds in body fluids [17], like ion chloride [18] or sweat quantity [19]. Furthermore, it is possible to directly functionalize fabrics with organic polymers in order to realize fully textile electrocardiogram electrodes for the detection of biopotentials [20,21]. Such devices represent a valid tool for the early diagnosis of diseases and monitoring. Wearable textile devices require that the electronic and data communication systems have to be flexible and lightweight to allow their integrability into the textiles themselves. This paper is focused on fully textile wearable pressure sensors. Many efforts have been devoted to developing sport garments and accessories to reduce injuries [22], increase the athlete's performance [23], comfort, or improve safety in the work place. Depending on the desired area of application, the pressure sensors have to be tuned to work in low or high pressure ranges with a reasonable sensitivity. Those with high sensitivity in the low pressure range (<10 kPa) allow touch detection in human-computer interaction or to develop an artificial hand for handling objects. In the same way, sensors working in a high pressure range (>10 kPa) can be used to monitoring the body pressure distribution in standing or sitting position. However, even if the linear response range and high sensitivity are the two main factors that describe a pressure sensor performance, it is necessary to consider its response related to a dynamic pressure change when it targets a real-life application. Very scarce and random details are reported in the literature on standard characterization procedures that would allow a comparison of the different proposed sensors' performances and actual fields of applicability.

Different examples of wearable pressure sensors are reported showing novel and simple configurations with high sensitivity, but most reported results have been obtained in a single step pressure measurement or in a very slow dynamic mode, that is, conditions that are far from the real-life applications. For instance, polyethylene combined with Carbon sheet has been applied to a glove and show good responses in a range between 1 kPa and 70 kPa. In this case, the sensor performance has been evaluated with a delay from 30 s to 180 s to reach rest conditions after each measure [24]. Another example is a piezoresistive sensor based on a modified graphite polyurethane that shows a slow dynamic range of almost 10 kPa/min [25]. On the contrary, good results have been obtained with a capacitive sensors [26], based on a graphene sponge with a wide response range (0–50 kPa), high sensibility of 0.96 kPa^{-1} and a good dynamic response demonstrated in a fast mode of 75 kPa/min. We recently developed a pressure sensor based on a conductive polymer printed on cotton fabric [22]. The poly (3,4-ethylenedioxythiophene): poly (4-styrenesulfonate) (PEDOT:PSS) is screen printed on the fabric and then two conductive threads are sewed to complete the electrical connections. This fabrication process is easy and low cost, with potential scalability for large-scale production. Even if the pressure sensor is rather simple on its own, its interaction with the textile nature of the substrate yields a complex system. The description of its working principle should take into account these challenging issues. We have recently reported an interpretation identifying three different contributions to textile pressure sensors performance [27]. At a macroscopic level the fabric properties, in terms of thickness, wettability and mechanical properties, are the most important factors that determine the pressure sensor performance. Going down to the microscopic level, the coating of the fibers affects the sensor response and sensitivity; finally, we identified a third contribution at the nano-level which influences the linear response range and the pressure saturation.

In this paper, we further deepen the investigation and the understanding of the mechanism controlling the performance of textile pressure sensors and present a detailed study of how the fabric type, weaving and structure influences their performance in dynamic operation mode. Four fabrics, with different material composition, thickness, sewing method and layered structure have been characterized and used as substrate for the pressure sensors. We performed electrical and mechanical tests (creep-recovery and stress-strain) both on plain fabrics and on the textile pressure sensors we fabricated onto each fabric, evaluating the sensor performance in dynamic mode operation in order to measure how the mechanical features of the fabrics affect this operation regime. We calculated and evaluated the sensor linear response in a wide pressure range. We used the same geometry and structure for all pressure sensors and we tested two different conducting polymer formulations—one based on pristine PEDOT:PSS and one obtained adding the ethylene glycol (EG)-like second dopant. We demonstrate that sensors based on the second formulation (more electrically conductive) show higher sensitivity, wider linear operation range but lower reliability compared to those realized with the first formulation (pristine solution). These results highlight the complexity of the textile pressure sensor system due to the active role of multiple parameters (i.e., the fabric composition, structure, the polymer formulation, the targeted pressure range, etc.) and suggest a protocol to optimize and tune textile pressure sensors to different types of application.

2. Materials and Methods

2.1. Materials

Four different fabrics, with different material composition, weave arrangement, layered structure and thickness were taken into account, considering both knitted and woven fabrics (Figure 1). The woven fabrics are all commercially available and are made by interlacing in a regular order two sets of yarns so that they cross each other perpendicularly. The lengthwise yarns are called warp while the crosswise yarns are called weft. Normally, they interlace each other at right angles. On the contrary, the knitted fabrics are made by interlocking a series of loops composed by one or more yarns, with every single row of loops locked into the previous row. The lengthwise loops are called wales and the crosswise loops are called courses.

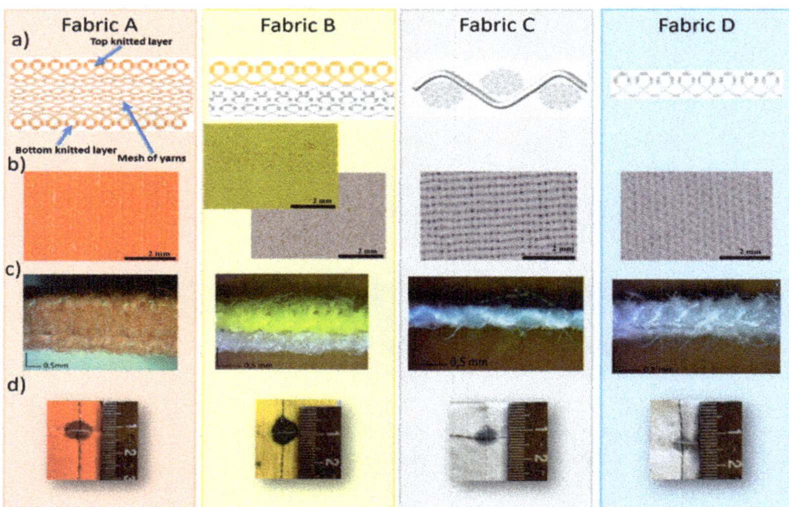

Figure 1. (**a**) Schematic representation of the fabric weaving structure; optical images of top (**b**) and lateral (**c**) view. (**d**) Appearance of one pressure sensor spot.

Fabric A consist of three layers with the outside ones composed by one single sheet of knitted structure. The embedded part is a system of yarns disposed perpendicularly with respect to the top and bottom sides. Fabric B is composed of two layers of knitted yarns distinguished by the different threads color. Fabric C, instead, is based on woven fabric. Finally, we also consider an elastic knitted textile, fabric D, whereby the special lengthwise stretchability is given by the particular knitting way of the yarns.

The conductive polymer poly (3,4-ethylenedioxythiophene): poly (4-styrenesulfonate (PEDOT:PSS), Clevious P, is purchased by Hearaeus. Two formulations of conductive ink have been studied—formulation Clevios P based on pristine PEDOT:PSS and formulation Clevios P+EG based on PEDOT:PSS with 10% v/v of ethylene glycol (supplied by Sigma-Aldrich).

The pressure sensors were prepared by drop casting the ink directly onto the fabric using a pipette. Fabrics A and B were drop cast with 25 µL in both sides, while the fabrics C and D were treated with 5 µL on one side. The conductive ink was dried at room temperature. A commercial 2 ply stainless steel conductive thread (resistivity of ~10 ohms/m) has been sewn with lockstitch by the sewing machine Elna eXa 320 in the final pressure sensor configuration as described elsewhere [27].

2.2. Fabric Chemical Characterization

Fourier transform-infrared (FT-IR) spectra were recorded on a Brucker Alpha Platinum-Attenuated Total Reflectance (ATR) spectrophotometer equipped with ATR Diamond window (32 scans, 4 cm^{-1} resolution).

Differential Scanning Calorimetry (DSC) measurements were carried out on a TA Instruments DSC Q2000 apparatus equipped with Refrigerated Cooling System RCS90, heating twice 3–5 mg samples in aluminum pans from −50 °C to 30 °C at 10 °C/min, with intermediate cooling run carried out at −10 °C/min. The instrument is calibrated with Indium standard.

2.3. Fabric Mechanical Characterization

TA Instrument Dynamic Mechanical Analysis (DMA) Q800 has been used in compression mode to characterize the mechanical properties of fabrics and sensors. The maximum force provided by DMA is 18 N, resulting in a normal pressure up to 100 kPa. In a creep recovery test, a constant pressure of 100 kPa was applied to the fabrics for 1 min and, after release, the strain recovery and thickness were recorded for three minutes. Stress-strain response has been evaluated with ten stress cycles from 0 kPa to 100 kPa at a speed of 50 N/min (~4.6 kPa/s).

2.4. Sensor Characterization

Pressure sensors performance has been studied using the DMA in compression mode together with the source meter Keithley 2400 as a read-out electronics. The same conditions described above for fabric characterization were used for the stress-strain tests. The compression top clamp has an area of 1.77 cm^2 (larger than the sensor active area) in order to provide an equal pressure distribution throughout the sensor. The source meter unit supplies a constant current of 1 μA, with a sampling time of 200 ms, meanwhile it measures the sensor resistance variations. This configuration allows monitoring the pressure, the thickness and the electrical behavior of the textile pressure sensors.

3. Results and Discussion

3.1. Characterization of Fabrics

All fabrics were analyzed in all of their components. Fabric A, an orange soft material, always displays a FT-IR pattern typical of nylon fibers (Figure 2), a synthetic material that is widely used for technical textiles [28]. While absorptions recorded in ATR mode account for a typical polyamide (i.e., nylon), with the 3293 cm^{-1} absorption ascribed to free N-H stretching in solid state, 1630 cm^{-1} band that is typical of carbonyl stretching (amide I band), and 1532 cm^{-1} signal ascribed to NH bending vibration (amide II band) [29] and 1, infrared spectroscopy is not helpful in identifying the exact composition of the polymer. Hence the thermal behavior of the fabric was analyzed in DSC, and the resulting thermogram in the second heating scan (Figure S1) displays two main events, both endothermic: the highest temperature signal centered at 252 °C that is typical of melting of the Nylon 6,6, crystal phase, the latter being a polyamide widely used in textile industry. The low temperature peak, which is persistent and not removed with the first heating scan intended for cancelling the thermal history of the sample, can instead be attributed to the melting of some waxy element applied as sizing coating to the fibers during the weaving processing.

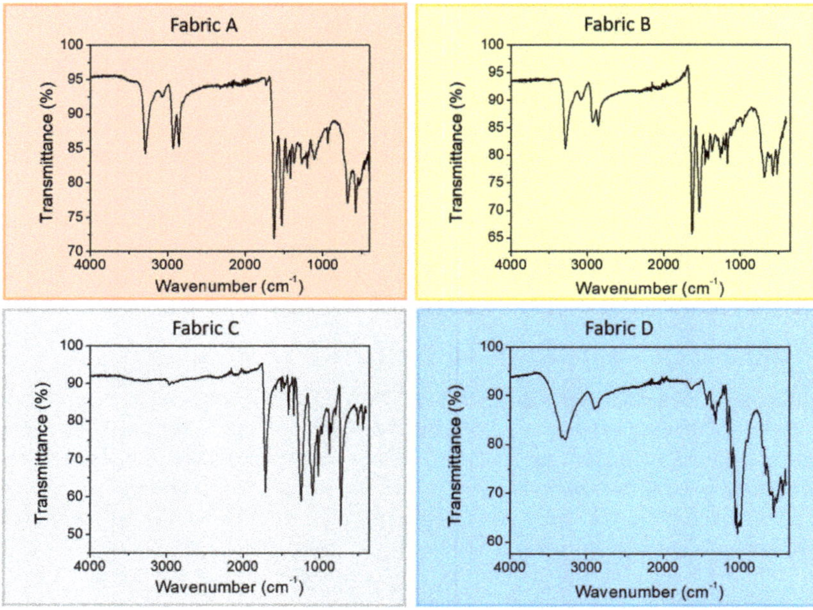

Figure 2. Fourier transform infrared (FT-IR) pattern for the four fabric studied.

Though fabric B displays a clear difference, at least in the coloring (white and yellow), of the two layers, when analyzed with the ATR-IR technique the sample displays exactly the same spectrum irrespective of the analyzed layer. As displayed in Figure 2, where the yellow layer spectrum is reported, the absorption pattern strongly resembles the previously discussed spectrum of fabric A, which was attributed to a polyamide fiber, identified as mainly Nylon 6,6. DSC analysis of fabric B, however, reveals a different thermal behavior than fabric A (Figure S1), with the presence of a high-T endothermic peak that is similar in position to the one attributed to the melting of Nylon 6,6, while a new endothermic event, positioned at a slightly lower T, is now present. The position of such a peak can be coherent with the melting of other commonly used polyamides, such as, among others, Nylon 6. Moreover, fabric B thermogram shows just the hint of a signal in the same position of the low-T endotherm observed for fabric A.

The fabric C ATR-IR spectrum displays a significantly different profile than the previously discussed ones (Figure 2), with features ascribed to a polyester filament. Indeed, the spectrum displays the major peaks associated with the structure of polyethylene terephthalate (PET) with the terephthalic acid ester carbonyl stretching at 1712 cm^{-1}, the asymmetric C–C–O and the O–C–C stretching at 1240 and 1091 cm^{-1}, respectively, and the C–H wagging vibrations from the aromatic structures at 722 cm^{-1} [30]. The DSC thermogram confirms this hypothesis, displaying a single endothermic event centered at 252 °C (Figure S1), that can be associated with the melting of PET crustal phase. Such a polyester is once again well renowned for its use in textiles and garments.

The last investigated sample, fabric D, shows an infrared absorption that is different from all the previously analyzed fibers (Figure 2), with the lack of any C = O feature, some prominent broad OH stretching band in the region 3500–3000 cm^{-1} and C-O-C signals in the 1150–1075 cm^{-1} region that are reminiscent of a cellulosic structure. Cellulosic fibers are the most relevant natural fibers used in textiles, both as cotton and flax fibers, suggesting that fabric D might be a cotton fabric. Cotton is not expected to provide any significant thermal signal in DSC analysis, since cellulosic fibers thermally degrade before being able to undergo any thermal transition. However, DSC second heating scan displays a low-T signal (Figure S1), in a position similar to that previously detected in fabric A and B.

In order to assess whether this is an additive, as previously hypothesized, fabric D was washed in water with soap, as a common garment. DSC analysis of the washed sample confirm the disappearance of the peak (Figure S2), supporting the hypothesis that this might be a processing coadjutant used in the weaving process.

3.2. Mechanical Characterization of Fabrics

We focused the structural analysis on the vertical properties of the fabrics because this is usually the relevant direction of the textile pressure sensors in real-life applications. To investigate and compare their properties we first performed a creep-recovery test. We selected the highest pressure used in this study (100 kPa) and we applied it for 1 min. The recovery curve was recorded for 3 min. This test provides information about the vertical elastic properties of each fabric and an influence on the textile pressure sensor performance is expected. Figure 3 shows the creep-recovery trends and the relevant parameters are summarized in Table 1.

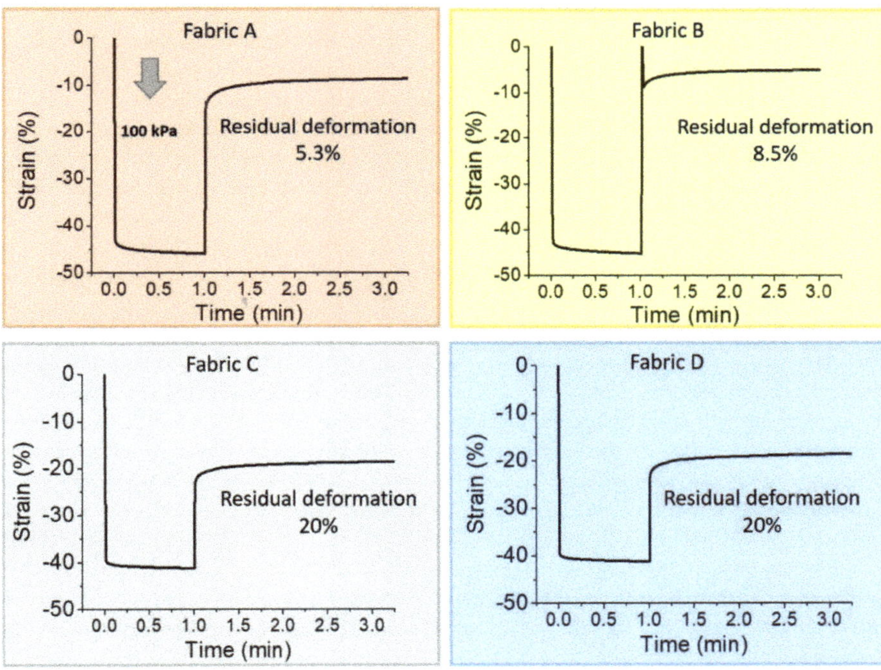

Figure 3. Mechanical characterization of the 4 different tested fabrics under creep recovery test: the arrow indicates the applied pressure (100 kPa for 1 min); the recovery time is 3 min.

Table 1. Summary of main parameters extracted from mechanical analyses.

	Creep Recovery				Stress-Strain Test
Fabric	**Initial Thickness (mm)**	**Thickness @100 kPa (mm)**	**Thickness after Release (mm)**	**Residual Deformation**	**Dynamic Strain Range**
A	2.49 *	1.34	2.36	5.3%	30%
B	1.54	0.74	1.40	8.5%	20%
C	0.36	0.19	0.29	20%	10%
D	0.71	0.37	0.57	20%	10%

* the DMA Q800 measures the thickness of the sample and the standard deviation associated to the average value correspond to a percentage error less than 1%.

The thickness of fabrics A and B are considerably different compared to the thickness of fabrics C and D. Strain is therefore the most appropriate quantity to be used in order to compare the creep-recovery test results. The maximum strain value reached by every fabric under the 100 kPa pressure was quite the same (~45%). Observing the residual deformation, fabric A and fabric B have good elastic properties with high recovery up to 5.3% and 8.5%, respectively. On the other hand, fabric C results are less elastic and its residual deformation is 20%. Comparting it with fabric D, we observed a similar behavior with the same recovery response. These results underline the fact that the elastic consistency of fabric D (the elastic one) mainly concern the horizontal stretching and does not affect the vertical mechanical properties.

Even though the creep recovery test gives relevant information about the fabric elastic properties, it does not represent their actual behavior in the textile pressure sensors during a dynamic operation mode. For this reason, we performed a cycle of compression and decompression stress-strain test. The stress-strain curves, in which the stress value represents the pressure, are reported in Figure 4. The fabrics show a substantial variation on the stress-strain curve after the first cycle. This is in agreement with the previous results, in which a complete recovery has not been observed for any fabric (Figure 3). The stress-strain test confirms that fabric A has good elastic properties, indicated by the linearity of the compression curve for each cycle. However, it does not have enough time to complete the recovery and after each cycle the strain curve does not overlap the previous one.

On the other hand, fabric B does not show proper elastic properties and the compression curve is not linear. However, apart for the first one, the strain curves are increasingly overlapping in subsequent cycles. Fabrics C and D do not show a linear compression curve. After the first cycle, both fabrics remain compressed but each strain curve almost overlaps that of the previous cycle. To better evaluate this aspect, the insets of Figure 4 show the strain against time (right axis) while cycled pressure is applied (left axis). In this representation, it is possible to observe the dynamic strain range (the difference between the maximum and the minimum strain value in a cycle) and the symmetries of the strain curve during compression and decompression. Ruling out the first cycle, in fabric A the strain value ranges from −10% to −40% for the second cycle, and from −13% to −43% for the tenth cycle, resulting in a large dynamic strain range of ~30%. As observed before, fabric B seems to be less elastic than fabric A but the strain values are more reproducible and range from −15% to −40% with a dynamic strain range of ~25%. Since the symmetry between compression and decompression curves, we can consider both fabrics provided with reasonable elastic features. The thinner fabrics behave differently, underlining the fact that the structure/geometry characteristics of textile can influence the response. They show a greater thickness variation after the first cycle and a lasting compression during the others. They have a lower dynamic strain range of about 10%. The strain values range from −35% to −45% for fabric C and from −40% to −30% for fabric D. Moreover, the compression and decompression regions of the strain curve are not symmetric and the smallest thickness does not correspond with the highest stress.

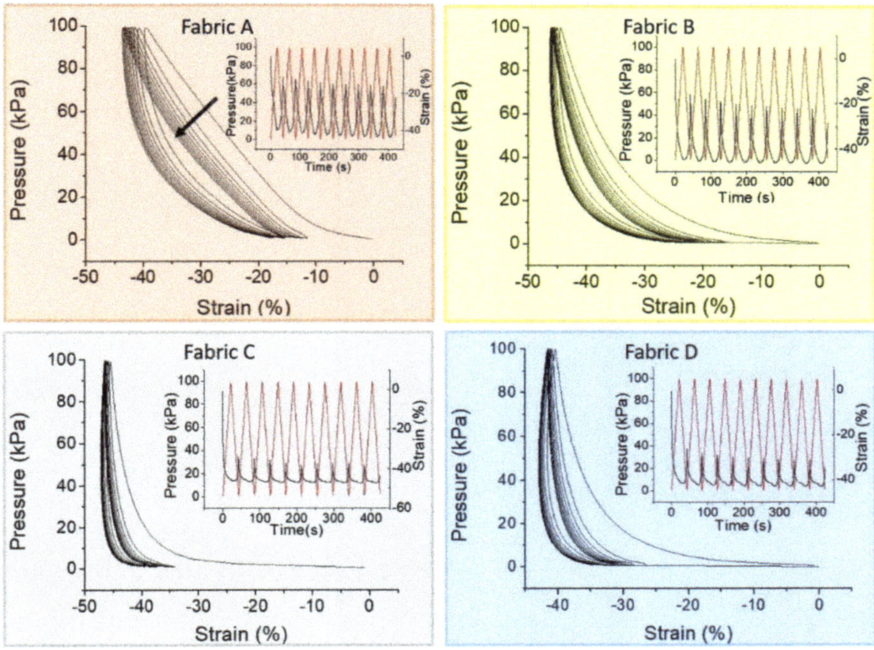

Figure 4. Stress-strain curve of fabrics: the arrow indicates the curve shift direction from the first to the tenth cycle. Inset: pressure (black line) and strain (red line) as a function of time for each fabric.

3.3. Pressure Sensor Performance in Dynamic Mode

We fabricated textile pressure sensors with each fabric, following the procedure described in the experimental method. The textile sensors, based on conductive polymer PEDOT:PSS, consist of two conductive stainless steel threads that are in contact with each other through a fabric coated with such a polymer. Increasing the applied pressure, the thickness of the fabric decreases, the piezoresistivity of conductive polymer changes and the whole pressure sensor resistance decreases (conductance increase) [22,27].

From now on, we will call sensor A the textile pressure sensor fabricated using fabric A, sensor B the one realized with fabric B and so on. Due to the differences in thickness, we increased the volume of the conductive polymer in the drop casting deposition in order to achieve a proper and comparable vertical coating also in thicker fabrics. For this reason, sensors A and B have a sensitive area (A = (51.6 ± 0.69) mm^2; B = (48.5 ± 0.6) mm^2) larger than the area of pressure sensors C and D (C = (33.2 ± 0.5) mm^2; D = (18.5 ± 0.4) mm^2). The upper graph of each section in Figure 5 shows the resistance response (black, on the left) during the application of 10 pressure cycles (red, on the right). The performance of the textile pressure sensor has been evaluated in the same condition as the stress-strain test. The resistance curve is inversely proportional to the applied pressure because, when the pressure increases, the device resistance decreases. A linear response of the electrical properties with the applied external pressure is more suitable for real-life applications. Thus, for our textile pressure sensor, it is useful to evaluate the conductance instead of the resistance. Plotting the conductance, a linear shape in some regions can be achieved. Figure 5c,d,g,h show the conductance trend under compression for each pressure cycle.

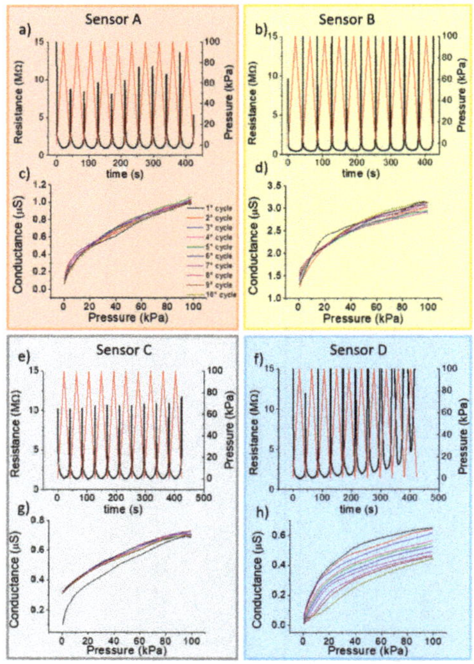

Figure 5. Comparison of the four fabrics. (**a**,**b**,**e**,**f**) Resistance sensor response in dynamic mode for 10 cycles with an applied pressure from 0 to 100 kPa with a frequency of 5 kPa/s. (**c**,**d**,**g**,**h**) Conductance trend during compression for every single cycle.

Sensors A and C, even if the fabrics used are substantially different, have a stable and reproducible response. Sensor C shows a strong variation after the first cycle because woven polyester fabric needs a first compression cycle to stabilize the vertical structure of its fibers. This behavior is also confirmed from the stress-strain test for this fabric. The conductance response for sensor B presents a small spread through 10 cycles that increases in the high pressure range. Sensor D, with a single layer of knitted cotton fabric, is the least stable because an increase in the resistance with the number of cycles is observed.

As already observed, the change of the conductive polymer formulations affects both its electrical and its piezoresistive properties and we assessed the effects of varying the conductive polymer formulation on the performance of textile pressure sensors. The consequences in the textile pressure sensors can be identified as a different performance and a different pressure operating ranges. We added to the PEDOT:PSS formulation 10% *v/v* of ethylene glycol which is a well know second dopant that enhance the conductivity of the organic compounds [31]. The cyclic conductance of the sensor response with the addition of EG is reported in Figure S3.

Figure 6 shows the average values of conductivity at each cycle with and without the presence of EG for textile pressure sensors fabricated with comparable geometry and structure onto the 4 different fabrics here tested. The linear working range, sensitivity and reliability are indeed affected by EG, the conductivity enhancement agent. The sensors based on the more conductive formulation have a greater sensitivity regardless of the fabric and a wider linear working range. However, the sensors with ethylene glycol present a less reliable behavior, as highlighted by larger standard deviation values. In these dynamic tests, we have not observed any saturation in the conductance as a function of pressure up to values of more than 70 kPa. This behavior is usually observed in a static characterization of the sensors realized with a more conductive ink in which the large number of conductive points between the PEDOT-rich zone that covers every single fiber enabling the achievement of a maximum conductance value before the application of the highest pressure.

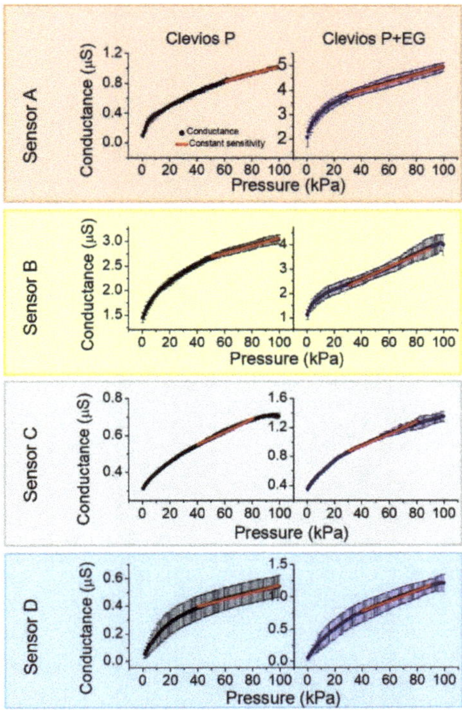

Figure 6. Average trend over ten cycles of the conductance response during compression stress for both formulations.

The linear conductance response (G) is modeled as $G = S{\cdot}P + G_{off}$, with S the sensitivity expressed in nS/kPa. G_{off} is the intercept value that is not zero because a preload force of 0.05 N is applied. Figure S4 shows the derivative of the conductance for every fabric in both formulations. The derivative represents the point sensitivity values and the constant region of these curves identify the sensors linear operation range. For a better estimation of this working range, the statistical χ^2-test was taken into account as descibed in the Supplementary Information.

The most relevant parameters to interpret the sensors performance are summarized in Table 2: (i) the conductance G_0 at the lowest applied pressure of 0.5 kPa; (ii) the conductance G_{100} when the pressure is 100 kPa and the sensitivity values calculated with the weighted least mean squares method. The sensitivity of a sensor is an essential parameter that describes how the estimated output varies when there is a variation in the input values. In this case, the output variable is the conductance which yields a different value according to the pressure variation (the input parameter). The sensitivity coefficient allows to know the effect on the measured output conductance due to a pressure change. The relation between the independent variable and the measured quantity is used as a calibration curve for developing the optimized sensor. The presence of an uncertainty in the conductance value during the dynamic mode operation gives information about the sensor reliability. The elastic behavior, the material of the fabric and the conductive formulation affect the sensor response after several working cycles. For example, the three-polyamide-layers (fabric A) and woven polyester fabrics (fabric C) are those that show the more stable response. The knitted-cotton based sensor (fabric D) is very unstable through the cycles and even if its sensitivity value calculated by analytical tools is comparable with that of polyester based sensors, it exhibits an excessive instability. In order to highlight the best materials and formulations for a real-life application of textile pressure sensors in a non-static framework, we computed the pressure uncertainty value δP_i in each single applied pressure as:

$$\delta P_i = \frac{\delta G_i}{\left[\frac{\partial G}{\partial P}\right]_{P = P_i}}.$$

Table 2. Main parameters of textile pressure sensors useful to evaluate their performances.

	Clevios P		Clevios P + EG	
Sensor	**G_0 (μS)**	**G_{100} (μS)**	**G_0 (μS)**	**G_{100} (μS)**
A	0.09 ± 0.02	1.01 ± 0.02	2.1 ± 0.4	4.4 ± 0.2
B	1.4 ± 0.1	3.03 ± 0.09	1.1 ± 0.2	4.0 ± 0.4
C	0.314 ± 0.003	0.70 ± 0.01	0.35 ± 0.03	1.35 ± 0.07
D	0.05 ± 0.03	0.54 ± 0.07	0.05 ± 0.03	1.2 ± 0.1
	S_{FIT} (nS/kPa)	**Linear Range**	**S_{FIT} (nS/kPa)**	**Linear Range**
A	4.90 ± 0.04	60–100 kPa	16.2 ± 0.1	30–100 kPa
B	7.70 ± 0.07	50–100 kPa	25.6 ± 0.2	30–90 kPa
C	3.47 ± 0.03	40–80 kPa	8.3 ± 0.1	30–80 kPa
D	2.54 ± 0.03	40–100 kPa	7.9 ± 0.1	40–90 kPa

Using this formula, we take into account the single point sensitivity $\frac{\delta G}{\delta P}_i$ calculated differentiating the curves in Figure 6, and the conductance uncertainty values δG_i This equation allows the extraction of information about the pressure uncertainty even if the sensitivity coefficient is not constant over all the studied pressure range. Figure 7 reports the δP values for each sensor using both conductive formulations. The large instability in the dynamic operation with a subsequent huge uncertainty in the pressure estimation forced us to reject fabric D as a candidate for real pressure sensors. The presence of etylhene glycol also gives a similar instability, with an associated relative error ≥30%. A possible reason could be that cotton fibers coated with the conducting polymer increase their stiffness and this can lead to a delamination or to cracks during the compression and decompression cycles, resulting in a variable trend. Similar results have been reported in the literature for similar mechanical stress levels [32]. The relative error associated to sensor A with EG, indicates how it is not reliable when the second dopant is present in the solution, even if it has an average better performance than sensor B without EG. Finally, the sensor based on polyester fabric (C), even if it is poorly elastic (either in vertical and horizontal direction), is the more reliable one, both with and without ethylene glycol. Fabric C is therefore the best candidate among those considered by us in order to realize pressure sensors able to monitor and record a dynamic pressure variation.

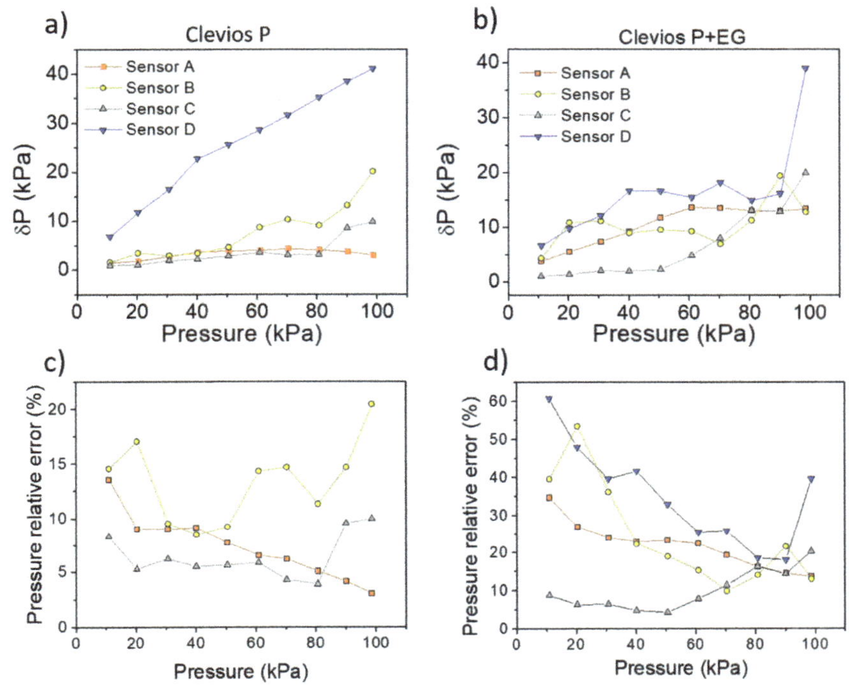

Figure 7. Pressure uncertainty values against the applied pressure (**a**,**b**). Here, for the sake of clarity, are reported the δP values only every 10 kPa. The inset in (**a**) also shows the trend of sensor D for the solution without ethylene glycol. The figures (**c**,**d**) report the pressure relative error for each sensor and for both formulations (excluding the sensor D with pristine formulation).

4. Conclusions

This work investigates the physical mechanisms controlling the performance of textile pressure sensors directly fabricated onto different fabrics realized with natural and synthetic fibers. It presents a detailed study of how the fabric type, weaving and structure influences the pressure sensors performance in dynamic operation mode. It demonstrates how the mechanical properties of the fabric substrates affect the textile pressure sensor fabricated onto them.

We studied the mechanical and sensing features of textile based pressure sensors directly fabricated onto four different types of textile substrates such as polyamide (A and B), polyester (C) and cotton (D), and with different structures.

The elastic properties of the fabric used to realize the textile pressure sensor resulted in having a relevant impact on the sensor performance. Fabrics with good elastic properties and wide dynamic strain range during compression result in a highly sensitive sensor (fabric A and B).

However, also fabrics with poor elastic features can be used to realize good textile pressure sensors that, indeed under compression respond with a lower strain dynamic range and thus have a limited sensitivity, but guarantee a high reliability and stability in dynamic operation mode (fabric C).

Textile pressure sensor A, based on a three-knitted layer of polyamide, shows good performance and high sensitivity in an high dynamic range (30%) due to its elastic behavior. Sensors based on fabric B show a lower reproducibility for each cycle due to the intermediate elastic properties of the pristine fabric. In this case, the macroscopic geometrical change during compression and decompression affects the sensor performance in terms of reliability even if it presents the highest sensitivity value. Sensor C, based on polyester, is the fabric with the worst elastic properties and it shows a very good stability, except for the first compression cycle. After the first compression, the thickness of fabric C varies under stress with a very low dynamic strain range (10%) that, however, results in a reasonable electrical output signal. Its sensitivity is lower with respect to sensor A, but it can still be very useful for real life applications. Fabric D, the elastic cotton fabric, shows similar mechanical properties along the vertical direction to fabric C but sensors D and C exhibit a very different behavior. In the case of sensor D, it may be the microscopic level which affects the performance and, as suggested from the conductance plots in Figure 7, after each cycle delamination or cracks occurring in the coating may lead to a deterioration of the PEDOT:PSS on the surface [29]. We thus conclude that textile sensors, fabricated onto fabrics with good elastic properties in the vertical direction, indeed show a greater sensitivity but could be less reliable if compared to similar sensors deposited onto a stiffer fabric.

We also assessed the effects of varying the conductive polymer formulation. Textile pressure sensors realized onto four fabric substrates (A, B, C and D) with a PEDOT:PSS formulation containing ethylene glycol show on average a higher sensitivity, associated with a lower stability and lower reliability with respect to comparable pressure sensors fabricated onto identical fabric substrates but with a pristine PEDOT:PSS formulation.

In summary, the results we report in this work underline how to understand and validate the performance of textile pressure sensors in a dynamic operation mode. The sensitivity is not the only parameter that must be taken into account and the sensor's response after multiple, repetitive compression cycles should also be considered. Fully textile pressure sensors are complex systems and in order to optimize them for real life applications it is important to consider several aspects, from the macroscopic to the microscopic scale. Reliability is an important and fundamental issue that should drive and guide the research and development of textile pressure sensors towards actual needs. We believe the results reported here open the way to a full optimization of textile pressure sensors in terms of most suitable fabrics, the range of linear pressure responses, sensitivity and dynamic operation behaviors to better fit the requirements of a large variety of dedicated real-life applications.

Supplementary Materials: The following are available online at http://www.mdpi.com/1424-8220/19/21/4686/s1, Figure S1: Differential Scanning Calorimetry (DSC) thermogram pattern for the four fabric., Figure S": DSC pattern for Fabric D (polyester) after washing, Figure S3: Compression cycles for textile pressure sensors realized with the addition in the conductive formulation of 10% *v/v* of ethylene glycol, Figure S4: Local sensitivity for each single sensor calculated differentiating the conductive curve, Figure S5. $\chi^2_{reduced}$ for the sensors based on Clevios P, Table S1: $\chi^2_{reduced}$ values for choose a linear operation range.

Author Contributions: Conceptualization, M.T. and B.F.; Data curation, L.P. and L.M.; Formal analysis, L.P. and L.M.; Investigation, L.P.; Methodology, M.T. and E.G.C.; Supervision, E.G.C. and B.F.; Writing–Original draft, L.P. and M.T.; Writing–Review & editing, E.G.C. and B.F.

Funding: This research was funded by project PO—MIUR ARS01_00996 "TEX-STYLE" (2019–2022).

Conflicts of Interest: The authors declare no conflict of interest.

References

1. Xu, S.; Zhang, Y.; Jia, L.; Mathewson, E.K.; Jang, K.I.; Kim, J.; Fu, H.; Huang, X.; Chava, P.; Wang, R.; et al. Soft microfluidic assemblies of sensors, circuits, and radios for the skin. *Science* **2014**, *344*, 70–74. [CrossRef]
2. Lee, H.; Song, C.; Hing, Y.S.; Kim, M.S.; Cho, H.R.; Kang, T.; Shin, K.; Choi, S.H.; Hyeon, T.; Kim, D.H.; et al. Wearable/disposable sweat-based glucose monitoring device with multistage transdermal drug delivery modul. *Sci. Adv.* **2017**, *3*, e1601314. [CrossRef] [PubMed]
3. Gao, W.; Emaminejad, S.; Yin, H.; Nyein, Y.; Challa, S.; Chen, K.; Peck, A.; Fahad, H.M.; Ota, H.; Shiraki, H.; et al. Fully integrated wearable sensor arrays for multiplexed in situ perspiration analysis. *Nature* **2016**, *529*, 509–514. [CrossRef] [PubMed]
4. Hwang, B.U.; Lee, J.H.; Trung, T.Q.; Roh, E.; Kim, D.I.; Kim, S.W.; Lee, N.E. Transparent stretchable self-powered patchable sensor platform with ultrasensitive recognition of human activities. *ACS Nano* **2015**, *9*, 8801–8810. [CrossRef] [PubMed]

5. Trindade, I.; Machado da Silva, J.; Miguel, R.; Pereira, M.; Luca, J.; Oliveira, L.; Valentim, B.; Barreto, J.; Santos Silva, M. Design and evaluation of novel textile wearable systems for the surveillance of vital signals. *Sensors* **2016**, *10*, 1573. [CrossRef]
6. Lamberti, F.; Sanna, A.; Rokne, J. Sensors for Entertainment. *Sensors* **2016**, *16*, 1102. [CrossRef]
7. Hwang, B.H.; Zabeeb, A.; Trung, T.Q.; Wen, L.; Lee, J.D.; Choi, Y.I.; Lee, H.B.; Kim, J.H.; Han, J.G.; Lee, N.E.L. A transparent stretchable sensor for distinguishable detection of touch and pressure by capacitive and piezoresistive signal transduction. *NPG Asia Mater.* **2019**, *11*, 23. [CrossRef]
8. Jacobs, J.V.; Hettinger, L.J.; Huang, Y.H.; Jeffries, S.; Lesch, M.F.; Simons, L.A.; Verma, S.K.; Willetts, J.L. Employee acceptance of wearable technology in the workplace. *Appl. Ergon.* **2019**, *78*, 148–156. [CrossRef]
9. Tessarolo, M.; Gualandi, I.; Fraboni, B. Recent progress in wearable fully textile chemical sensors. *Adv. Mater. Technol.* **2018**, *3*, 1700310. [CrossRef]
10. Carvalho, H.; Yao, Y.; Gonçalves, L.M. Flexible force sensors for e-textiles. In *IOP Conference Series: Materials Science and Engineering*; IOP Publishing: Bristol, UK, 2017; Volume 254, p. 072007.
11. Zhou, B.; Sundholm, M.; Cheng, J.; Cruz, H.; Lukowicz, P. Measuring muscle activities during gym exercises with textile pressure mapping sensors. *Pervasive Mob. Comput.* **2017**, *38*, 331–345. [CrossRef]
12. Lin, X.; Seet, B.-C. Battery-Free Smart Sock for Abnormal Relative Plantar Pressure Monitoring. *IEEE Trans. Biomed. Circuits Syst.* **2017**, *11*, 464–473. [CrossRef] [PubMed]
13. Büscher, G.H.; Kõiva, R.; Schürmann, C.; Haschke, R.; Ritter, H.J. Flexible and stretchable fabric-based tactile sensor. *Rob. Auton. Syst.* **2015**, *63*, 244–252. [CrossRef]
14. Kirthika, S.K.; Ponraj, G.; Ren, H. Fabrication and Comparative Study on Sensing Characteristics of Soft Textile-Layered Tactile Sensors. *IEEE Sens. Lett.* **2017**, *1*, 1–4. [CrossRef]
15. Maziz, A.; Concas, A.; Khaldi, A.; Stålhand, J.; Persson, N.K.; Jager, E.W.H. Knitting and weaving artificial muscles. *Sci. Adv.* **2017**, *3*, e1600327. [CrossRef]
16. Kinkeldei, T.; Zysset, C.; Cherenack, K.H.; Troster, G. A textile integrated sensor system for monitoring humidity and temperature. In Proceedings of the 2011 16th International Solid-State Sensors, Actuators and Microsystems Conference, Beijing, China, 5–9 June 2011; pp. 1156–1159.
17. Gualandi, I.; Marzocchi, M.; Achilli, A.; Cavedale, D.; Bonfiglio, A.; Fraboni, B. Textile Organic Electrochemical Transistors as a Platform for Wearable Biosensors. *Sci. Rep.* **2016**, *6*, 33637. [CrossRef]
18. Gualandi, I.; Tessarolo, M.; Mariani, F.; Cramer, T.; Tonelli, D.; Scaveta, E.; Fraboni, B. Nanoparticle gated semiconducting polymer for a new generation of electrochemical sensors. *Sens. Actuators B Chem.* **2018**, *273*, 834–841. [CrossRef]
19. Jia, J.; Xu, C.; Pan, S.; Xia, S.; Wei, S.; Noh, H.; Zhang, P.; Jiang, X. Conductive Thread-Based Textile Sensor for Continuous Perspiration Level Monitoring. *Sensors* **2018**, *18*, 3775. [CrossRef]
20. Pani, D.; Dessi, A.; Gusai, E.; Saenz-Cogollo, J.F.; Barabino, G.; Fraboni, B.; Bonfiglio, M. Evaluation of novel textile electrodes for ECG signals monitoring based on PEDOT:PSS-treated woven fabrics. In Proceedings of the 2015 37th Annual International Conference of the IEEE Engineering in Medicine and Biology Society (EMBC), Milan, Italy, 25–29 August 2015; pp. 3197–3200.
21. Castrillón, R.; Pérez, J.J.; Andrade-Caicedo, H. Electrical performance of PEDOT: PSS-based textile electrodes for wearable ECG monitoring A comparative study. *Biomed. Eng. Online* **2018**, *17*, 38. [CrossRef]
22. Saenz-Cogollo, J.; Pau, M.; Fraboni, B.; Bonfiglio, A. Pressure Mapping Mat for Tele-Home Care Applications. *Sensors* **2016**, *16*, 365. [CrossRef]
23. Meyer, J.; Lukowicz, P.; Troster, G. Textile Pressure Sensor for Muscle Activity and Motion Detection. In Proceedings of the 2006 10th IEEE International Symposium on Wearable Computers, Montreux, Switzerland, 11–14 October 2006; pp. 69–72.
24. Pizarro, F.; Villavicencio, P.; Yunge, D.; Rodriguez, M.; Hermosilla, G.; Leive, A. Easy-to-Build Textile Pressure Sensor. *Sensors* **2018**, *18*, 1190. [CrossRef]
25. He, Y.; Li, W.; Yang, G.; Liu, G.; Lu, J.; Zheng, T.; Li, X.; He, Y.; Li, W.; Yang, G.; et al. A novel method for fabricating wearable, piezoresistive, and pressure sensors based on modified-graphite/polyurethane composite films. *Materials* **2017**, *10*, 684. [CrossRef] [PubMed]
26. Tian, H.; Shu, Y.; Wang, X.F.; Mohammad, M.A.; Bie, Z.; Xie, Q.Y.; Li, C.; Mi, W.T.; Yang, Y.; Ren, T.L. A graphene-based resistive pressure sensor with record-high sensitivity in a wide pressure range. *Sci. Rep.* **2015**, *5*, 8603. [CrossRef] [PubMed]

27. Tessarolo, M.; Possanzini, L.; Campari, E.G.; Bonfiglioli, R.; Violante, F.S.; Bonfiglio, A.; Fraboni, B. Adaptable pressure textile sensors based on a conductive polymer. *Flex. Print. Electron.* **2018**, *3*, 034001. [CrossRef]
28. Byrnes, C.; Horrocks, A.R.; Anand editors, S.C. Technical Textile Market–An Overview. In *Handbook of Technical Textiles*; CRC Press: Boca Raton, FL, USA, 2000.
29. Silverstein, R.M.; Bassler, G.C.; Morrill, T.C. *Spectrometric Identification of Organic Compounds*, 7th ed.; John Wiley and Sons Inc.: New York, NY, USA, 2005.
30. Strain, I.N.; Wu, Q.; Pourrahimi, A.M.; Hedenqvist, M.S.; Olsson, R.T.; Andersson, R.L. Electrospinning of recycled PET to generate tough mesomorphic fibre membranes for smoke filtration. *J. Mater. Chem. A* **2015**, *3*, 1632–1640. [CrossRef]
31. Zhang, S.; Kumar, P.; Nouas, A.S.; Fontaine, L.; Tang, H.; Cicoira, F. Solvent-induced changes in PEDOT: PSS films for organic electrochemical transistors. *APL Mater.* **2015**, *3*, 014911. [CrossRef]
32. Del Agua, I.; Mantione, D.; Ismailov, U.; Sanchez-Sanchex, A.; Aramburu, N.; Malliaras, G.G.; Mecerreyes, D.; Ismailova, E. DVS-Crosslinked PEDOT:PSS Free-Standing and Textile Electrodes toward Wearable Health Monitoring. *Adv. Mater. Technol.* **2018**, *3*, 1700322. [CrossRef]

sensors

MDPI

Article

Comparative Performance of Four Electrodes for Measuring the Electromechanical Response of Self-Damage Detecting Concrete under Tensile Load

Hyeon Woo Noh, Min Kyoung Kim and Dong Joo Kim *

Department of Civil and Environmental Engineering, Sejong University, 209, Neungdong-ro, Gwangjin-gu, Seoul 05006, Korea

* Correspondence: djkim75@sejong.ac.kr; Tel.: +82-2-3408-3820; Fax: +82-2-3408-4332

Received: 23 July 2019; Accepted: 19 August 2019; Published: 21 August 2019

Abstract: Self-damage or/and stress-sensing concrete is a promising area of research for measuring the electromechanical response of structural materials using more robust sensors. However, the copper and silver paste sensors widely used in such applications can be expensive and have detrimental effects on the load carrying capacity and durability of the structural systems upon which they are installed. Accordingly, this study compared the performance of four electrode types—conventional copper tape with silver paste (CS), copper film with type 1 carbon tape (CC1), copper film with type 2 carbon tape (CC2), and copper wire and film with type 2 carbon tape (WC2)—to develop an economical and practical electrode for measuring the electromechanical response of self-damage-detecting concrete. The CC1 electrode exhibited comparable performance to the CS electrode in measuring the electromechanical response of self-damage-detecting concrete, despite requiring a longer polarization time (80 s) than the CS electrode (25 s). The CS electrode exhibited a higher damage-sensing capacity (GF_2), whereas the CC1 electrode exhibited a higher strain-sensing capacity (GF_1), as well as good damage-sensing capacity. Therefore, the CC1 electrode using copper film with type 1 carbon tape was determined to be the best alternative to the conventional CS electrode.

Keywords: self-sensing; attached electrode; self-damage detecting concrete; copper material; carbon material

1. Introduction

Catastrophic collapses of civil infrastructure and buildings have brought about increased attention to the necessity of structural health monitoring (SHM) systems [1–3]. Current SHM systems mostly utilize embedded and/or attached sensors [4–6], but these types of sensors have exhibited relatively low durability and very limited sensing capacity. Accordingly, over the last two decades, much research has been performed on the self-sensing capacity of cement-based construction materials containing conductive fillers, such as steel fibers, carbon fibers, multi-walled carbon nanotubes, carbon black, and graphite powder under various loads [7–23]. The electrical impedance or resistivity of self-sensing concrete containing electrically conductive fillers is well-known to change under stress and/or due to the occurrence of damage, including cracks. In this application, it is important to correctly measure the pure electrical response of the self-sensing concrete under load.

The correct evaluation of damage or stress based on the electromechanical response of self-damage and/or stress-sensing concrete requires suitable electrodes that have the following properties recommended by Han et al. [24] and Azhari [25]: electrode materials should be electrically conductive and durable under severe environmental exposure and repeated mechanical loads; the addition of the electrode should not negatively affect the concrete strength; and the electrode should be easily connected to a data acquisition system. Azhari [25] recommended copper and silver paste as suitable electrode materials. Indeed, many researchers have used copper and silver paste in electrodes when investigating the electro-mechanical responses of various self-sensing concretes [13,15,17,26–30]. However, current embedded-type electrodes utilizing copper materials can possibly generate negative effects on both the load-carrying capacity and durability of the structural systems that contain them. Additionally, attached-type electrodes utilizing silver paste are very expensive and are not reusable. A suitable electrode should be highly conductive, affordable, and easily attached to the surface of self-damage and/or stress-sensing concrete.

The objective of this study is accordingly to determine a suitable electrode for the measurement of the electromechanical response of self-damage detecting concrete by utilizing or combining common electrically conductive materials. The specific objectives are (1) to investigate the electrical polarization depending on the type of electrode, and (2) to measure the electromechanical response of self-damage-detecting concrete under tensile load using the evaluated electrodes.

2. Current Electrodes

2.1. Embedded-Type Electrodes

Current embedded-type electrodes mostly include steel or copper wire mesh or wire gauze consisting of grid structures. Han and Ou [30] embedded copper wire meshes in self-sensing concrete containing carbon-based fillers to measure the electrical response under repeated compressive loads. They reported an 8% decrease in the electrical resistivity of self-sensing concrete containing these carbon-based fillers. Sun et al. [31] utilized steel wire meshes as embedded electrodes to measure the electromechanical responses of self-sensing concretes containing graphite. As the magnitude of the applied compressive load increased, the electrical resistivity of the graphite-reinforced self-sensing concretes decreased—the reduction in electrical resistivity was 15.6%. Both steel and copper wire meshes demonstrated good performance as embedded type electrodes. Azhari [25] investigated the effects of curing time on the electrical resistivity of cement-based materials by using different electrodes (copper plate, copper mesh, and copper wire with silver paste)—the embedded-type electrodes (copper plate and copper mesh) produced a smaller change in the measured electrical resistivity than the attached-type electrode (copper wire with silver paste).

2.2. Attached-Type Electrodes

Current attached-type electrodes generally utilize silver paste, silver paint, copper wire, or copper tape. Song et al. [15] applied an attached-type electrode using copper tape and silver paste on the surface of high performance fiber-reinforced cementitious composites (HPFRCCs) to measure their electromechanical response during direct tensile tests. They reported that the electrical resistivity of the HPFRCCs clearly decreased as the number of micro-cracks increased within the gauge length. Azhari and Banthia [13] also used attached-type electrodes with copper wire and silver paste to measure the electromechanical response of self-sensing concretes containing carbon materials, which exhibited a noticeable decrease (25%) in the electrical resistivity under repeated compressive load.

Han et al. [24] compared the electrical polarization of an embedded-type electrode (copper gauze) and an attached-type electrode (copper foil). The embedded electrode generated a lower polarization potential than the attached electrode, although both electrodes produced different polarization potentials corresponding to different amounts of current. They reported that the polarization potential was affected by the different effective areas between the electrode and the matrix—copper foil with a larger effective area exhibited higher polarization potential than copper gauze with a lower effective area. Thus, different electrodes have been observed to produce different electrical and/or electromechanical responses in cement-based composites according to their different geometric and material properties.

3. Experimental

The experimental program in this study, described in Figure 1, was designed to compare the performance of four different electrodes when measuring the electrical response of self-damage detecting concrete. The configuration of the electrode can be divided into two components: the probe component, which measures the electrical impedance or resistance of the self-damage detecting concrete, and the adhesive component, which transfers the applied electrical current into the self-damage-detecting concrete. Materials evaluated in this study included copper film and copper wire, and two different types of carbon tape were evaluated as the adhesive component.

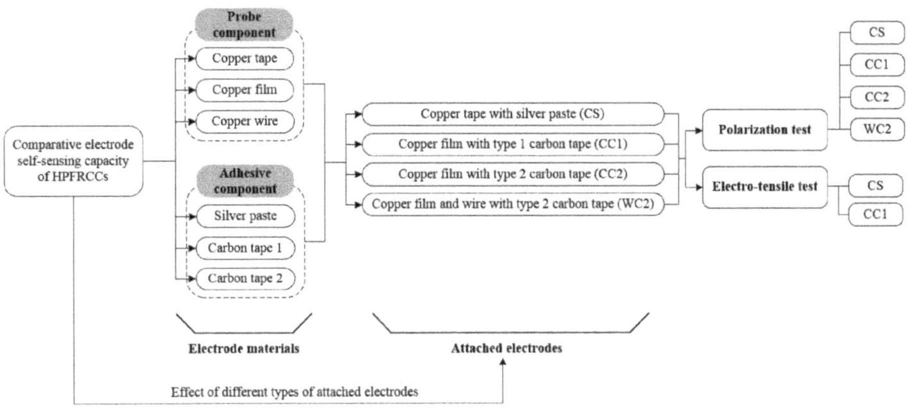

Figure 1. Experimental program. HPFRCCs = high performance fiber-reinforced cementitious composites.

Table 1 summarizes the properties of the electrode materials evaluated in this study. By combining different materials in the probe and adhesive components, three types of electrode were investigated as follows: 1) copper film with type 1 carbon tape (CC1); 2) copper film with type 2 carbon tape (CC2); and 3) copper wire and film with type 2 carbon tape (WC2). A conventional electrode consisting of copper tape with silver paste (CS) was also evaluated as a control. The performance of each electrode was evaluated by measuring the polarization effect and by investigating the electromechanical response of the HPFRCCs under tension.

Table 1. Properties of electrode materials.

Electrode Material	Thickness, t (mm)	Width, w (mm)	Length, l (mm)	Cross-Section, A (mm^2) †	Electrical Resistivity ($\times 10^{-7}$ kΩ·cm)
Copper tape	0.12	10	100	1.20	0.058 [a]
Copper film	0.30	5	50	1.50	0.023 [b]
Carbon tape type 1	0.16	10	100	1.60	1.3 [a]
Carbon tape type 2	0.18	5	100	0.90	1.3 [a]
Copper wire	0.16 *		1000	0.02	0.000017 [a]

† A = t × w, * diameter of copper wire, [a] manufacturer obtained the measured electrical resistivity, [b] measured electrical resistivity using the four probe method.

3.1. Materials and Specimen Preparation

Specimens of HPFRCC containing 1 vol% long and 1 vol% medium smooth steel fibers were investigated to evaluate the performance of the three electrodes (CC1, CC2, and WC2). The composition of the matrix and its compressive strength are provided in Table 2. The diameter of the silica sand was 0.36 mm on average and the water-to-cement ratio was 0.35. The compressive strength of the matrix was measured to be 95 MPa by 100 × 200 mm cylinder specimen tests. The properties of the steel fibers are summarized in Table 3—the length and diameter of the long smooth steel fibers were 30 and 0.3 mm, respectively, while those of the medium smooth steel fibers were 19.5 and 0.2 mm, respectively.

Table 2. Matrix composition and compressive strength.

Cement (Type III)	Silica Sand (#40) *	Fly Ash	Super-Plasticizer **	Water	Compressive Strength (MPa)
1.0	1.0	0.15	0.0055	0.35	95

* average particle size of silica sand is 0.355 mm, ** super-plasticizer contained 25% solid content.

Table 3. Properties of steel fibers.

Fiber Type	Diameter (mm)	Length (mm)	Density (g/cc)	Tensile Strength (MPa)	Elastic Modulus (GPa)	Electrical Resistivity ($\times 10^{-10}$ kΩ·cm) *
Long smooth	0.3	30	7.90	2447	200	6.9
Medium smooth	0.2	19.5	7.90	2942	200	5.4

* measured electrical resistivity using four probe method.

A 20 L capacity Hobart-type laboratory mixer was used to mix the specimens. Cement, silica sand, and fly ash were first dry mixed for 10 min. Then, water was added and further mixed for 5 min. When the mortar mixture exhibited suitable workability and viscosity to provide uniform fiber distribution, the shorter fibers were first carefully dispersed, then the longer fibers were added in by hand. The mortar mixture with the fibers was then further mixed for 1–2 min.

The mortar mixture containing the steel fibers was poured into molds for both the polarization and tensile test specimens, as can be seen in Figure 2a. Two layers of steel wire mesh were used to reinforce both ends of the specimen, as can be seen in Figure 2b. The specimens were then covered with a plastic sheet and kept in a laboratory at room temperature for 24 h. Next, the specimens were demolded and cured in water for 2 weeks. After curing, the specimens were kept in dry conditions for 2 h, then the surfaces of the specimens were ground to accommodate the attachment of electrodes. At least three specimens (SP1, SP2, and SP3) were prepared for each electrode test series.

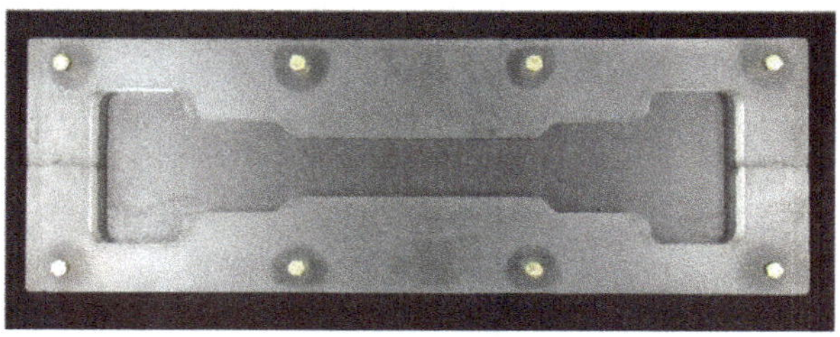

(a) mold for tensile specimen

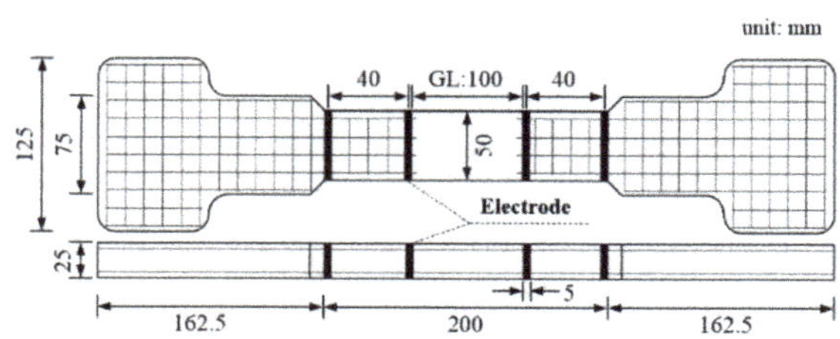

(b) geometry of tensile specimen

Figure 2. Tensile specimen mold and geometry.

3.2. Test Set-Up and Procedure

The four-probe measurement method proposed by Wenner [32], which requires two outer input current electrodes and two inner output current electrodes, was employed to measure the electrical polarization and electromechanical response of the self-damage-detecting concrete, as shown in Figure 3. The distance between the two inner electrodes was 100 mm, while that between the inner and outer electrodes was 40 mm. The geometry of tensile specimens and test set-up in this study were referred to relevant previous studies [17,33].

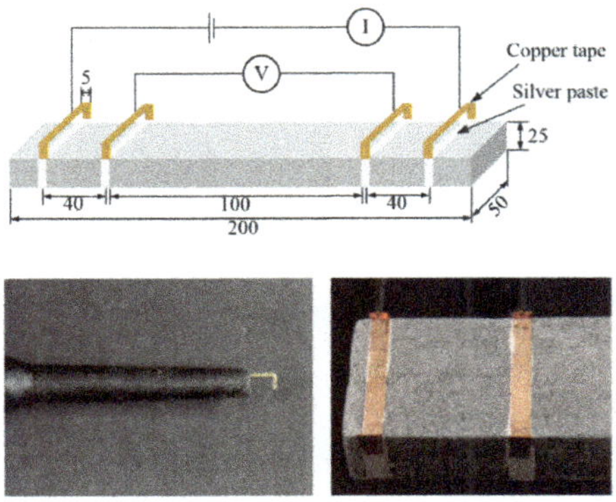

(a) CS (silver paste)

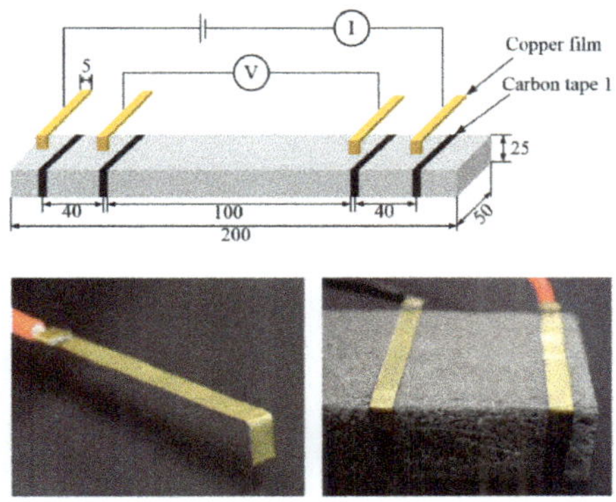

(b) CC1 (copper film with type 1 carbon tape)

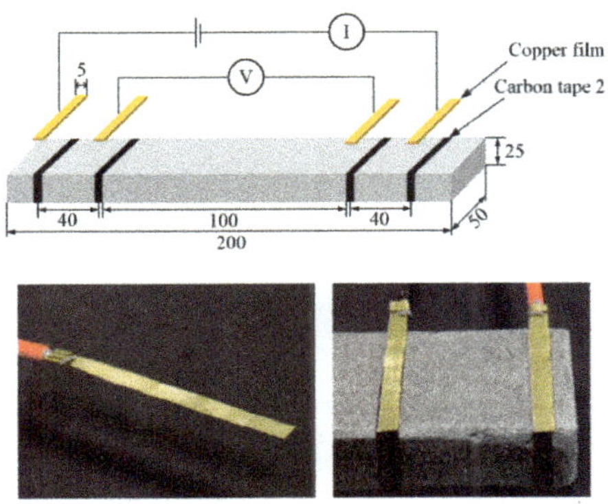

(c) CC2 (copper film with type 2 carbon tape)

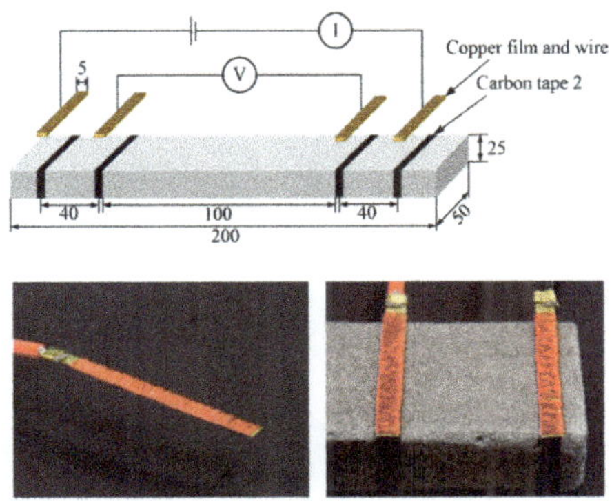

(d) WC2 (copper wire and film with type 2 carbon tape)

Figure 3. Electrode arrangement with different types of electrodes.

A commercial multimeter (Keysight 3458A, Santa Rosa, California, USA) was used to measure the change in electrical resistance during the tests. The electrical resistivity of the HPFRCCs was measured by using direct current (DC) measurements to obtain a more accurate electrical resistivity [34]. The magnitude of the input current was maintained at 50 μA for 14 min to monitor the electrical polarization. During the polarization tests, the specimens were kept in a chamber with a constant temperature (25 °C) and humidity (60%). The test setup is illustrated schematically in Figure 4.

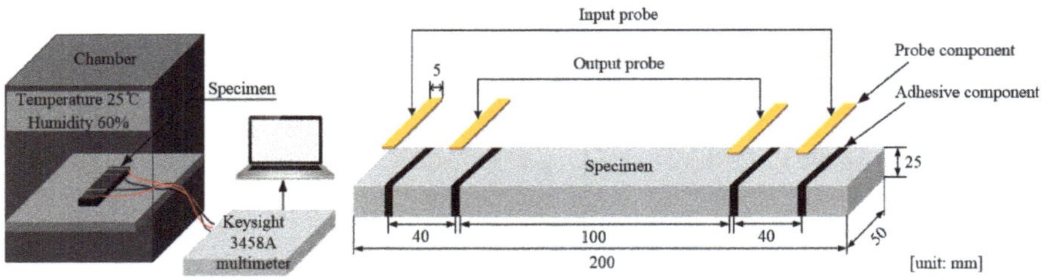

Figure 4. Polarization effect stabilization test set-up and details about the specimen.

Figure 5 shows the test set-up for measuring the electrical resistance of the HPFRCCs during the direct tensile tests. A universal testing machine (UTM) was used to apply a constant 1.0 mm/min rate of displacement. The tensile elongation of the specimens was measured by two linear variable differential transformers (LVDTs) installed in an aluminum cage, while the tensile stress was measured by a 50 kN capacity load cell. To measure the electrical resistance of the specimens under tension, the Keysight 3458A multimeter was connected to the specimen, as shown in Figure 5. Prior to tensioning, the electrical resistance was measured for at least 15 min to stabilize the effects of electrical polarization.

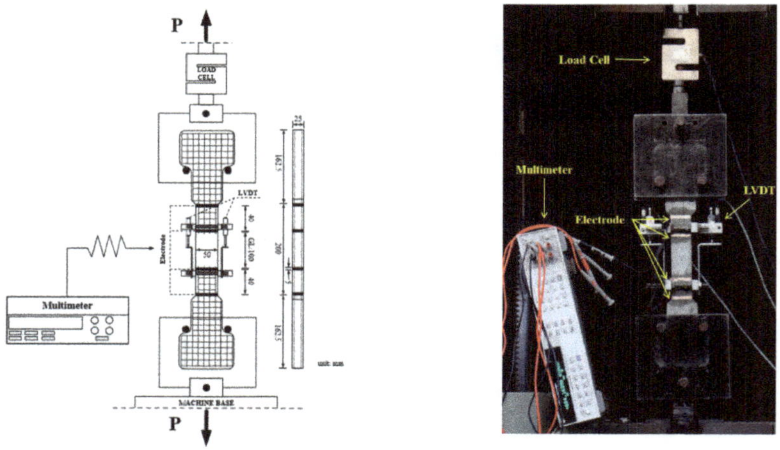

Figure 5. Set-up for measuring the electrical resistance of HPFRCCs during direct tensile tests. LVDT = linear variable differential transformers.

4. Results and Discussion

Figure 6a,b shows the typical change in electrical resistivity of the HPFRCCs due to electrical polarization before applying tensile load and under direct tension, respectively. In Figure 6a, under electric current without load, the electrical resistivity can be observed to rapidly increase until reaching the polarization time required for stable electrical resistivity. In Figure 6b, it can be observed that as the tensile strain (ε) increased from 0 to the first cracking strain (ε_{cc}), the tensile stress (σ) linearly increased from 0 to the first cracking strength (σ_{cc}, point A) of the HPFRCCs under direct tension, while the electrical resistivity (ρ) decreased from the initial electrical resistivity (ρ_0) to that at the first cracking point A′ (ρ_{cc}). Even though the slight change in electrical resistance within the elastic limit prior to points A and A′ cannot be clearly correlated to the change in applied tensile stress, the electrical resistivities of the HPFRCCs beyond points A and A′ significantly decreased until the post-cracking points B and B′, as illustrated in Figure 6b. The electrical resistivity considerably decreased from ρ_{cc} (point A′) to the post-cracking electrical resistivity at point B′ (ρ_{pc}), while the tensile strain increased from ε_{cc} to the post-cracking strain (ε_{pc}), and the tensile stress increased from σ_{cc} to the post-cracking strength (σ_{pc}, point B). Kim et al. [17] reported that the electrical resistance of HPFRCCs under direct tension with multiple micro-cracks is composed of a non-cracked component and a cracked component. Since the electrical conductivity of the steel fibers in the cracked portion of an HPFRCC is much higher than that in the non-cracked portion, the electrical resistance decreases as the number of micro-cracks increases.

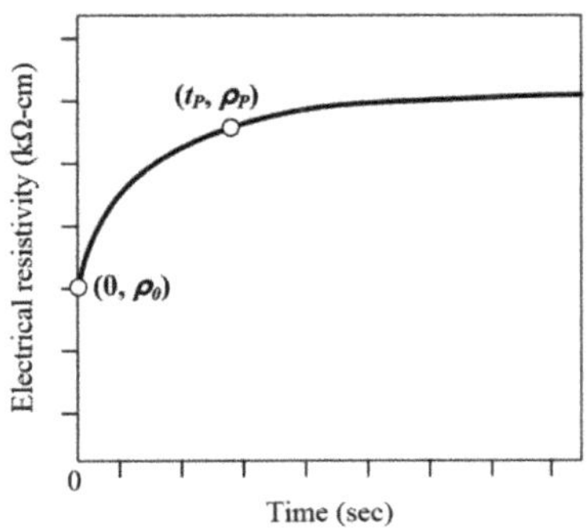

(a) under electric current without load

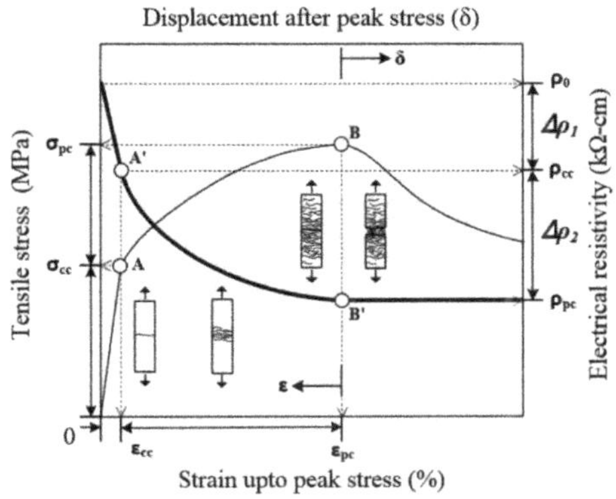

(b) under direct tensile load

Figure 6. Typical change of electrical resistivity of HPFRCCs.

The electrical resistivity (ρ) was calculated using the following Equation (1):

$$\rho = R\frac{A}{L}, \tag{1}$$

where ρ is electrical resistivity (kΩ·cm), R is the electrical resistance (kΩ), A is the cross-sectional area of the specimen (cm^2), and L is the distance between the input and output electrodes (cm).

Figure 7 shows the electrical resistivity history of identical HPFRCCs for 840 s (14 min) prior to loading, measured using the four evaluated electrode types (CS, CC1, CC2, and WC2) under electric current without any applied load. The multimeter was calibrated to collect data only within a range of 6 to 0.1 Hz to minimize noise. The electrical responses measured using the CS, CC1, and CC2 electrodes clearly show the typical electrical polarization phenomena for 840 s, as can be seen in Figure 7, while the response measured using the WC2 electrode did not. Among the three electrodes that exhibited a typical polarization response, the CS electrode exhibited the shortest polarization time (25 s). Based on the polarization tests, CC2 and WC2 electrodes were excluded in the investigation of the electromechanical response of HPFRCCs under direct tensile load, as both electrodes produced significant polarization and electrical noise.

Figure 8 shows the electromechanical responses of the HPFRCCs under direct tensile load as measured using electrodes types CS and CC1, both of which exhibit the typical electromechanical response of HPFRCCs with steel fibers. To quantify the strain- and damage-sensing capacity of the HPFRCCs, their gauge factors (GF, GF_1, and GF_2) were calculated using the following Equation (2) [17]:

$$GF = \frac{\Delta\rho}{\rho_0 \cdot \varepsilon_{pc}};\ GF_1 = \frac{\Delta\rho_1}{\rho_0 \cdot \varepsilon_{cc}};\ GF_2 = \frac{\Delta\rho_2}{\rho_{cc} \cdot (\varepsilon_{cc} - \varepsilon_{pc})}, \tag{2}$$

where GF_1 represents the strain-sensing capacity within the elastic range prior to first cracking, GF_2 denotes the damage-sensing capacity of the HPFRCC from after first cracking to the post-cracking point, and the overall sensing-capacity can be estimated using GF. Specimen SP1 of the HPFRCC using the CS electrode was excluded from the calculation of the average GF values due to the large deviation of its test results.

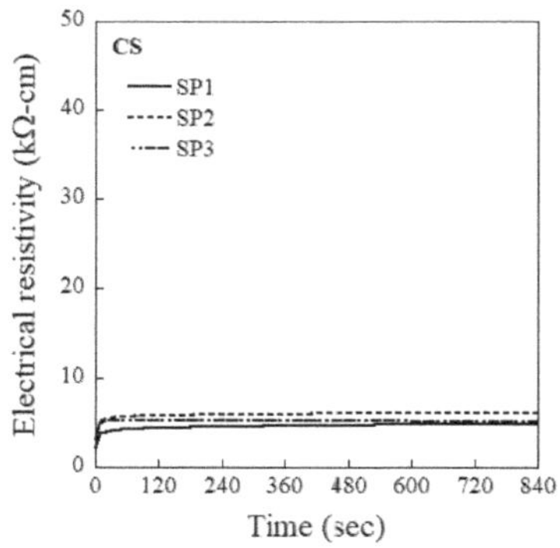

(a) CS

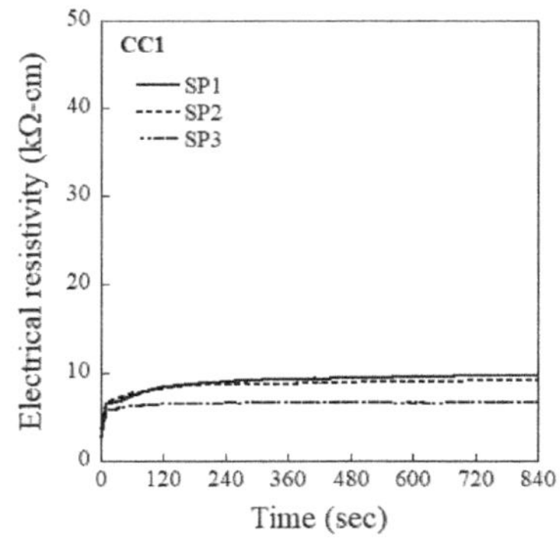

(b) CC1

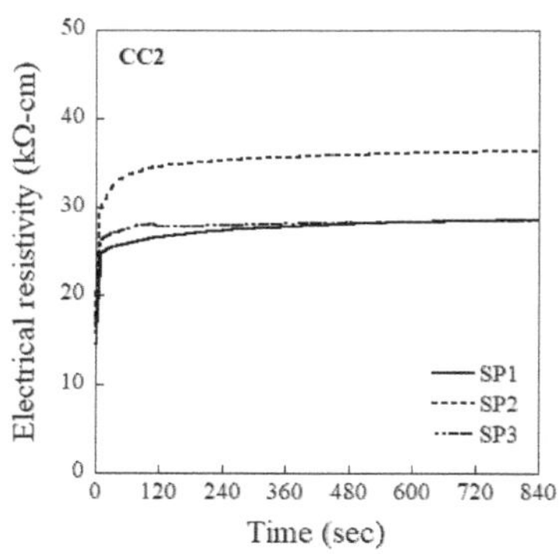

(c) CC2

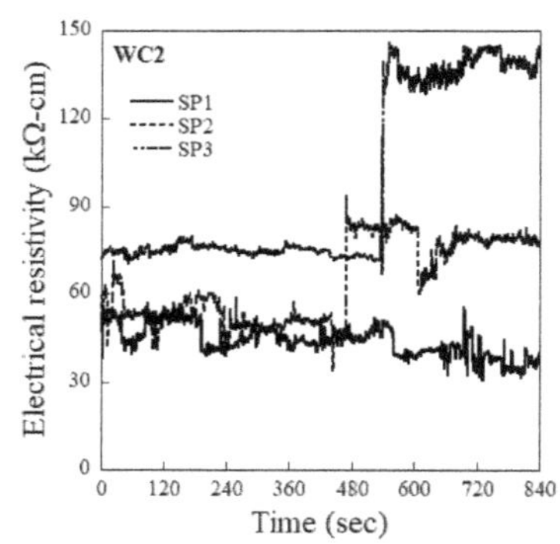

(d) WC2

Figure 7. Electrical resistivity history with time for HPFRCCs with different types of electrode (calibrated 0.1 Hz).

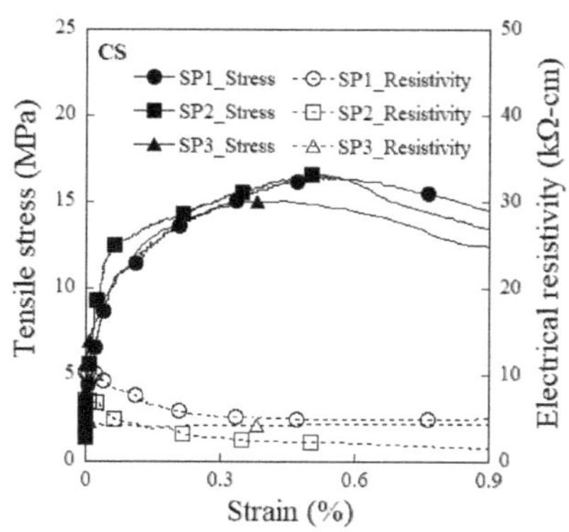

(a) CS

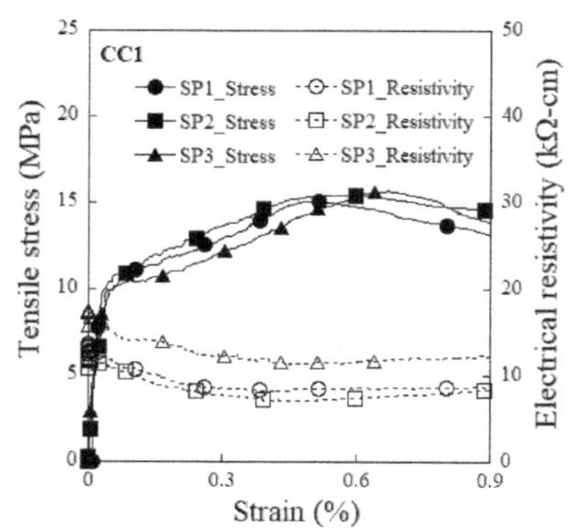

(b) CC1

Figure 8. Electro-tensile behavior of HPFRCCs with different types of electrodes.

The HPFRCCs using the CC1 electrode exhibited a higher strain-sensing capacity ($GF_1 = 2.39$) than those using the CS electrode, as well as good damage-sensing capacity ($GF_2 = 0.65$). Consequently, the electromechanical response of the HPFRCCs measured using the CC1 electrode generated comparable self-strain and damage-sensing capacity to that measured using the conventional CS electrode.

4.1. Effects of Different Electrodes on the Polarization Parameters

Table 4 summarizes the parameters describing the electrical polarization corresponding to the four different evaluated electrodes. These polarization parameters consist of the initial electrical resistivity (ρ_0), electrical resistivity at polarization time (ρ_P), fractional change in the electrical resistivity from ρ_0 to ρ_P ($\overline{\rho_p}$), change in the electrical resistivity at polarization time ($\Delta\rho_P$), slope at polarization time (ρ'_P), and polarization time (t_p). Figure 9 shows the measured change in the electrical resistivity ($\Delta\rho$) and the variation in the slope (ρ') of that change corresponding to the different types of electrodes evaluated, used to determine the polarization time (t_p) required to establish stable electrical resistivity, defined in this study as the time satisfying both of the following conditions: 1) when $\Delta\rho$ is less than 0.09 kΩ·cm, and 2) when ρ' is less than 0.009 kΩ·cm/sec. To minimize noise when determining t_p, the data measured at a frequency of 6 Hz was calibrated to the data measured at a frequency of 0.1 Hz, as can be seen in Figure 9a,b. The above conditions for determining t_p could then be obtained using the measured data calibrated to 0.1 Hz. As a result, the t_p for the specimens equipped with the CS, CC1, and CC2 electrodes was determined to be 25 s, 80 s, and 107 s, respectively.

Figure 10 shows the relationships between the fractional change in the electrical resistivity and polarization time of the CS-, CC1-, and CC2-equipped HPFRCCs. The fractional changes in the electrical resistivity at the polarization time ($\overline{\rho_p}$) of the CS-, CC1-, and CC2-equipped HPFRCCs were 186.7%, 240.7%, and 190.5%, respectively. The value of $\overline{\rho_p}$ can be observed to be closely related to the value of t_p—the $\overline{\rho_p}$ of the CS- and CC1-equipped HPFRCCs increased from 186.7% to 240.7%, as the t_p increased from 25 s to 80 s.

Table 4. Polarization response parameters of HPFRCCs with different types of electrodes. SP = specimen.

Type of Electrode	No.	Electrical Resistivity (kΩ·cm)			Fractional Change in Electrical Resistivity (%)	Slope at Polarization Time (kΩ·cm/s) *	Polarization Time (s)
		ρ_0	ρ_P	$\Delta\rho_P$	$\overline{\rho_p}$	$d\Delta\rho_P/dt_p$	t_p
CS	SP1	2.2	4.0	0.10	190.8	0.0020	30
	SP2	3.2	5.5	0.12	173.8	0.0020	30
	SP3	2.7	5.3	0.06	195.5	0.0040	20
	Avg. [a]	2.7	4.9	0.09	186.7	0.0027	25
	STD [b]	0.4	0.7	0.02	9.3	0.0009	4.7
CC1	SP1	2.8	8.1	0.13	295.8	0.0010	110
	SP2	3.8	8.1	0.08	211.3	0.0000	90
	SP3	2.8	6.4	0.04	214.9	0.0000	40
	Avg.	3.2	7.5	0.08	240.7	0.0000	80
	STD	0.5	0.8	0.04	39.0	0.0005	29.4
CC2	SP1	14.6	26.5	0.07	180.3	0.0010	100
	SP2	16.5	34.6	0.11	210.6	0.0010	130
	SP3	15.5	27.7	0.04	180.5	0.0060	90
	Avg.	15.5	29.6	0.07	190.5	0.0027	107
	STD	0.8	3.6	0.03	14.2	0.0024	17.0
WC2	SP1	51.0					
	SP2	48.8					
	SP3	71.3			-		
	Avg.	57.0					
	STD	10.1					

* absolute value of the calculated value, [a] Avg.: average values; [b] STD: standard deviation.

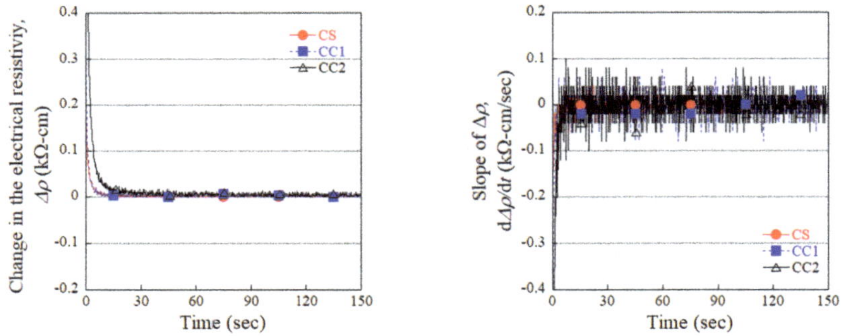

(**a**) 6 Hz

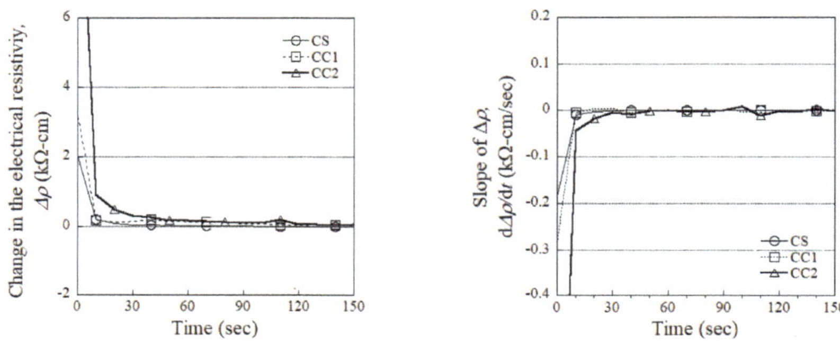

(**b**) 0.1 Hz

Figure 9. Change in the electrical resistivity and slope of change in the electrical resistivity with calibrated different data frequency.

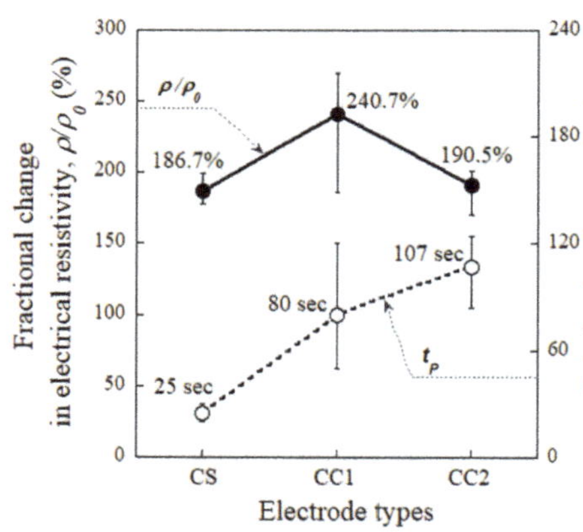

Figure 10. Relationship between the fractional change in the electrical resistivity and polarization time of HPFRCCs corresponding to the electrode types.

The correlation between $\overline{\rho_p}$ and t_p, describing the polarization phenomenon, can be explained by the electron flow at the interface between the electrode probes and the HPFRCCs. Suryanto et al. [35] reported that the electrical resistance of cementitious composites was affected not only by the matrix characteristics (compressive strength, temperature, humidity, etc.) but also the electrical resistance of the interface. Figure 11a–c illustrates the electron accumulation phenomenon at the interface between each electrode type and the HPFRCCs. When the input current (i) flows from the probe into the cement based material, negative electrons (e^-) move in the opposite direction [36]. At this time, electron accumulation, which causes the polarization effect, occurs between the probe (which has a high conductivity) and the cement based material (which has a low conductivity). As the electron accumulation increases, both the polarization time (t_p) and the fractional change in the electrical resistivity at polarization time ($\overline{\rho_p}$) increase. Therefore, as can be seen in Figure 11a, electrons at the interface between the CS electrode and the HPFRCC, which exhibited a shorter t_p, accumulated less than for other electrodes, whereas the electrons at the interface between the CC2 electrode and the HPFRCC, which exhibited a longer t_p, accumulated more, as can be seen in Figure 11c. Consequently, the $\overline{\rho_p}$ and t_p of the CC1-equipped HPFRCC, which exhibits more electron accumulation between the specimen and the electrode than the CS-equipped HPFRCC, were higher than those of the CS-equipped HPFRCC. The results obtained using the CC2 electrode showed the highest t_p (107 s), and a $\overline{\rho_p}$ (190.5%) lower than that when using CC1 electrode, as the CC2 electrode (which used a smaller-area carbon tape as the adhesive) seemed to generate significantly more electron accumulation at the interface between the specimen and the electrode. Thus, the CC2 electrode was determined to be unsuitable for measuring the electrical resistance of HPFRCCs. Among the evaluated electrodes, the CS electrode exhibited the shortest t_p (25 s). Among the remaining electrodes, because electrode CC1 exhibited a shorter t_p (80 s) than electrode CC2 (107 s), CC1 was determined to be more suitable than CC2 for measuring electrical polarization. Accordingly, the CS and CC1 electrodes were selected as the focus of the remaining investigation.

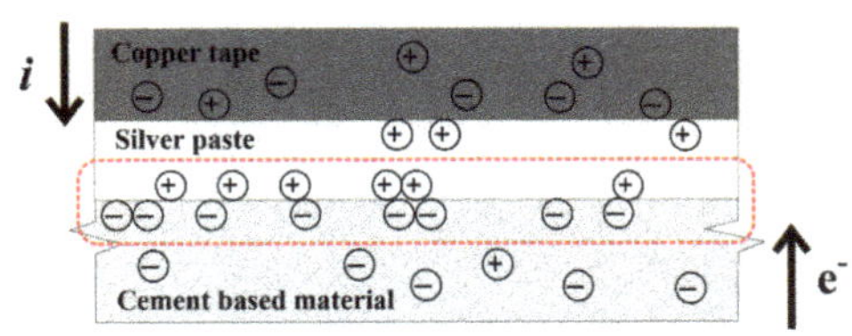

(**a**) using the CS electrode

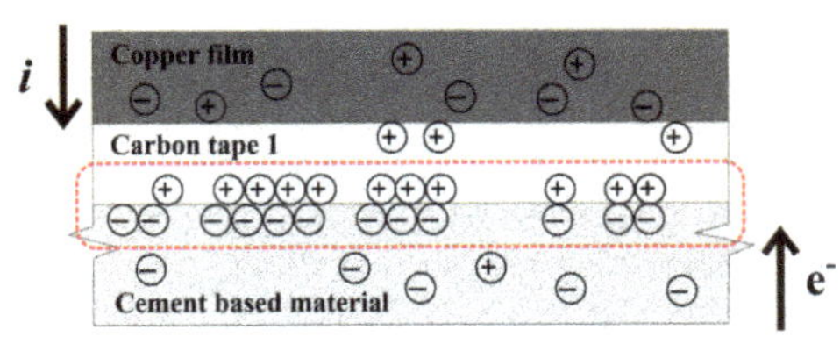

(**b**) using the CC1 electrode

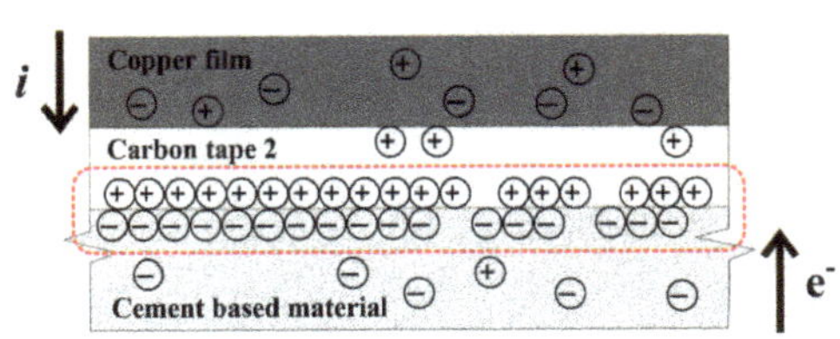

(**c**) using the CC2 electrode

Figure 11. Accumulation phenomenon of electrons at the interface between the probe and HPFRCCs.

4.2. Effects of Electrode Types on Electromechanical Response

Table 5 summarizes the electromechanical response parameters (ε_{cc}, σ_{cc}, ε_{pc}, σ_{pc}, ρ_0, ρ_{cc}, and ρ_{pc}) of HPFRCCs corresponding to the two most effective electrode types evaluated in this study (CS and CC1). All specimens, regardless of the type of electrode used, exhibited a tensile strain-hardening response and similar mechanical resistance. The average first-cracking strength (σ_{cc}) was 5.5 MPa and 6.4 MPa for the CS- and CC1-equipped HPFRCCs, respectively; the average post-cracking strength (σ_{pc}) was 14.9 MPa and 15.7 MPa, respectively; and the average strain capacity (ε_{pc}) was 0.48% and 0.56%, respectively.

Table 5. Electro-tensile parameters of HPFRCCs with CS and CC1 electrodes.

Notation		Tensile Strain (%)		Tensile Stress (MPa)		Electrical Resistivity (kΩ-cm)			Change in the Electrical Resistivity (kΩ-cm)		
		ε_{cc}	ε_{pc}	σ_{cc}	σ_{pc}	ρ_0	ρ_{cc}	ρ_{pc}	$\Delta\rho$	$\Delta\rho_1$	$\Delta\rho_2$
CS	SP1	0.01	0.44	6.0	13.3	5.1	4.3	4.2	0.9	0.8	0.1
	SP2	0.02	0.48	6.0	15.8	7.1	6.9	2.2	4.9	0.2	4.7
	SP3	0.02	0.51	4.5	15.4	10.4	10.2	4.9	5.6	0.3	5.3
	Avg.	0.02	0.48	5.5	14.9	8.8	8.5	3.5	5.3	0.3	5.0
	STD	0.00	0.02	0.75	0.20	1.65	1.65	1.35	0.35	0.05	0.30
CC1	SP1	0.03	0.50	5.3	15.1	12.8	12.7	8.5	5.2	1.0	4.2
	SP2	0.02	0.54	6.4	15.5	12.1	11.4	7.3	4.8	0.6	4.2
	SP3	0.03	0.65	7.4	15.7	17.4	16.4	11.9	5.6	1.1	5.6
	Avg.	0.03	0.56	6.4	15.4	14.1	13.5	9.2	4.9	0.6	4.3
	STD	0.00	0.06	0.86	0.25	2.35	2.12	1.95	0.33	0.22	0.66

There was clear difference in the measured initial electrical resistivity (ρ_0) of the CS- and CC1-equipped HPFRCCs, which were 8.8 kΩ·cm and 14.1 kΩ·cm, respectively. However, as can be seen in Figure 12, the overall change in electrical resistivity ($\Delta\rho$) was 5.3 kΩ·cm and 4.9 kΩ·cm for the CS- and CC1-equipped HPFRCCs, respectively, even though the materials used in the CS electrode exhibited a higher electrical conductivity than those used in the CC1 electrode.

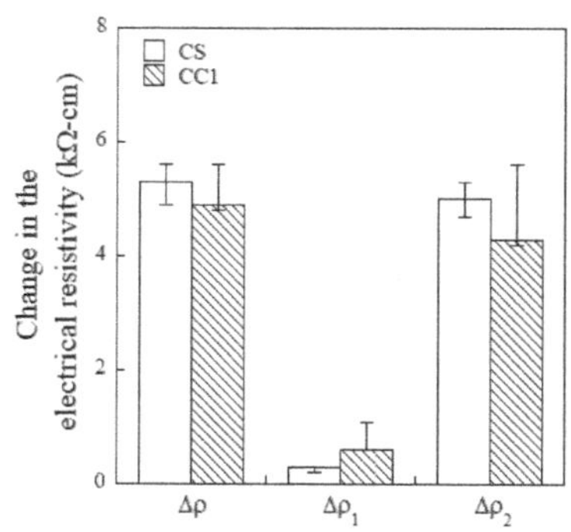

Figure 12. Change in the electrical resistivity of HPFRCCs with CS and CC1 electrodes under direct tensile load.

Table 6 compares the effects of electrode type on the gauge factors. The CS-equipped HPFRCCs showed a higher damage-sensing capacity (GF_2), whereas the CC1-equipped HPFRCCs exhibited a higher strain-sensing capacity (GF_1), as well as good damage-sensing capacity—GF_1 of the CC1-equipped HPFRCC was 2.39 and its GF_2 was 0.65. Therefore, the CC1 electrode using copper film with type 1 carbon tape was identified as the best suited for use as a replacement for the conventional copper tape with silver paste (CS) electrode in measuring the electromechanical response of HPFRCCs.

Table 6. Gauge factors of HPFRCCs with CS and CC1 electrodes under direct tensile load.

Notation		Overall Sensing Capacity, GF	Strain-Sensing Capacity, GF_1	Damage-Sensing Capacity, GF_2
CS	SP1 [¥]	0.40	15.69	0.05
	SP2	1.44	1.41	1.48
	SP3	1.06	1.44	1.06
	Avg.	1.22	1.43	1.23
	STD	0.19	0.02	0.21
CC1	SP1	0.81	2.60	0.70
	SP2	0.73	2.48	0.71
	SP3	0.50	2.11	0.56
	Avg.	0.62	2.39	0.65
	STD	0.13	0.21	0.07

[¥] the specimen 1 (SP1) of the HPFRCCs using the CS electrode was excluded in the calculation of the average.

5. Conclusions

In this study, we investigated the electromechanical response performance of four electrodes for use in self-damage detecting concrete: a conventional copper tape with silver paste (CS), copper film with type 1 carbon tape (CC1), copper film with type 2 carbon tape (CC2), and copper wire and copper film with type 2 carbon tape (WC2). The CC1 electrode exhibited the best performance among the three types of electrodes investigated to replace the conventional CS electrode. On the basis of the experimental results, the following conclusions can be drawn:

- Both the CC1 and CS electrodes produced a relatively shorter polarization time than the other electrodes (CC2 and WC2). The polarization times (t_p) for the specimens using the CS and CC1 electrodes were 25 s and 80 s, respectively.
- The t_p was significantly affected by the accumulation of electrons at the interface between the specimen and electrode. Higher levels of electron accumulation resulted in longer polarization times.
- The overall change in the electrical resistivity ($\Delta\rho$) upon damage to specimens equipped with the CS and CC1 electrodes was 5.3 kΩ·cm and 4.9 kΩ·cm, respectively. The CS-equipped HPFRCCs showed a higher damage-sensing capacity (GF_2), whereas the CC1-equipped HPFRCCs exhibited a higher strain-sensing capacity (GF_1), as well as good damage-sensing capacity.
- The CC1 electrode using copper film with type 1 carbon tape is identified as the best suited for use as a replacement for the conventional copper tape with silver paste (CS) electrode in measuring the electromechanical response of self-damage-detecting concrete.

We determined that the CC1 electrode can be used as a replacement for the conventional copper tape with silver paste (CS) electrode under direct tensile load. This is an improvement in economic terms as well, as using carbon tape is a third of the cost of using silver paste. In future research, we intend to demonstrate the suitability of the CC1 electrode for use with HPFRCCs under compressive load, cyclic load, and other measuring conditions (using alternating current, biphasic direct current, multiple electrodes, the two probe method, reusability and accounting for temperature influence, etc.) to more completely capture the capabilities and performance of this electrode type.

Author Contributions: Investigation, H.W.N. and M.K.K.; writing—review and editing, D.J.K.

Funding: This research was supported by a grant (19CTAP-C143065-02) from the Infrastructure and Transportation Technology Promotion Research Program funded by the Ministry of Land, Infrastructure and Transport of the Korean government.

Acknowledgments: This research was supported by a grant (19CTAP-C143065-02) from the Infrastructure and Transportation Technology Promotion Research Program funded by the Ministry of Land, Infrastructure and Transport of the Korean government.

Conflicts of Interest: The authors declare no conflict of interest.

References

1. Xu, D.Y.; Banerjee, S.; Wang, Y.B.; Huang, S.F.; Cheng, X. Temperature and loading effects of embedded smart piezoelectric sensor for health monitoring of concrete structures. *Constr. Build. Mater.* **2015**, *76*, 187–193. [CrossRef]
2. Perry, M.; Saafi, M.; Fusiek, G.; Niewczas, P. Hybrid optical-fibre/geopolymer sensors for structural health monitoring of concrete structures. *Smart Mater. Struct.* **2015**, *24*, 045011. [CrossRef]
3. Li, W.J.; Kong, Q.Z.; Ho, S.C.M.; Lim, I.; Mo, Y.L.; Song, G.B. Feasibility study of using smart aggregates as embedded acoustic emission sensors for health monitoring of concrete structures. *Smart Mater. Struct.* **2016**, *25*, 115031. [CrossRef]
4. Morsy, R.; Marzouk, H.; Haddara, M.; Gu, X. Multi-channel random decrement smart sensing system for concrete bridge girders damage location identification. *Eng. Struct.* **2017**, *143*, 469–476. [CrossRef]
5. Yazdizadeh, Z.; Marzouk, H.; Hadianfard, M.A. Monitoring of concrete shrinkage and creep using Fiber Bragg Grating sensors. *Constr. Build. Mater.* **2017**, *137*, 505–512. [CrossRef]
6. Liu, P.; Wang, W.; Chen, Y.; Feng, X.; Miao, L.X. Concrete damage diagnosis using electromechanical impedance technique. *Constr. Build. Mater.* **2017**, *136*, 450–455. [CrossRef]
7. Banthia, N.; Djeridane, S.; Pigeon, M. Electrical resistivity of carbon and steel micro-fiber reinforced cements. *Cem. Concr. Res.* **1992**, *22*, 804–814. [CrossRef]
8. Fu, X.; Lu, W.; Chung, D.D.L. Improving the strain-sensing ability of carbon fiber-reinforced cement by ozone treatment of the fibers. *Cem. Concr. Res.* **1998**, *28*, 183–187. [CrossRef]
9. Bontea, D.M.; Chung, D.D.L.; Lee, G.C. Damage in carbon fiber-reinforced concrete, monitored by electrical resistance measurement. *Cem. Concr. Res.* **2000**, *30*, 651–659. [CrossRef]
10. Chen, B.; Wu, K.; Yao, W. Conductivity of carbon fiber reinforced cement-based composites. *Cem. Concr. Compos.* **2004**, *26*, 291–297. [CrossRef]
11. Wang, W.; Dai, H.Z.; Wu, S.G. Mechanical behavior and electrical property of CFRC-strengthened RC beams under fatigue and monotonic loading. *Mater. Sci. Eng. A* **2008**, *479*, 191–196. [CrossRef]
12. Chung, D.D.L. Carbon materials for structural self-sensing, electromagnetic shielding and thermal interfacing. *Carbon* **2012**, *50*, 3342–3353. [CrossRef]
13. Azhari, F.; Banthia, N. Cement-based sensors with carbon fibers and carbon nanotubes for piezoresistive sensing. *Cem. Concr. Compos.* **2012**, *34*, 866–873. [CrossRef]
14. Han, B.G.; Zhang, L.Q.; Sun, S.W.; Yu, X.; Dong, X.F.; Wu, T.J.; Ou, J.P. Electrostatic self-assembled carbon nanotube/nano carbon black composite fillers reinforced cement-based materials with multifunctionality. *Compos. Part A* **2015**, *79*, 103–115. [CrossRef]
15. Song, J.D.; Nguyen, D.L.; Manathamsombat, C.; Kim, D.J. Effect of fiber volume content on electromechanical behavior of strain-hardening steel-fiber-reinforced cementitious composites. *Compos. Mater.* **2015**, *49*, 3621–3634. [CrossRef]
16. Nguyen, D.L.; Song, J.D.; Manathamsombat, C.; Kim, D.J. Comparative electromechanical damage-sensing behaviors of six strain-hardening steel fiber-reinforced cementitious composites under direct tension. *Compos. Part B* **2015**, *69*, 159–168. [CrossRef]
17. Kim, M.K.; Kim, D.J.; An, Y.K. Electro-mechanical self-sensing response of ultra-high-performance fiber-reinforced concrete in tension. *Compos. Part B* **2018**, *134*, 254–264. [CrossRef]
18. Materazzi, A.L.; Ubertini, F.; D'Alessandro, A. Carbon nanotube cement-based transducers for dynamic sensing of strain. *Cem. Concr. Compos.* **2013**, *37*, 2–11. [CrossRef]
19. Pisello, A.L.; D'Alessandro, A.; Sambuco, S.; Rallini, M.; Ubertini, F.; Asdrubali, F.; Materizzi, A.L.; Cotana, F. Multipurpose experimental characterization of smart nanocomposite cement-based materials for thermal-energy efficiency and strain-sensing capability. *Sol. Energy Mater. Sol. Cells* **2017**, *161*, 77–88. [CrossRef]
20. Hou, T.C.; Lynch, J.P. Electrical impedance tomographic methods for sensing strain fields and crack damage in cementitious structures. *J. Intell. Mater. Syst. Struct.* **2008**, *20*, 1363–1379. [CrossRef]
21. Hallaji, M.; Seppanen, A.; Pour-Ghaz, M. Electrical impedance tomography-based sensing skin for quantitative imaging of damage in concrete. *Smart. Mater. Struct.* **2014**, *23*, 085001. [CrossRef]
22. Gupta, S.; Gonzalez, J.G.; Loh, K.J. Self-sensing concrete enabled by nano-engineered interfaces. *Struct. Health. Monit.* **2016**, *16*, 309–323. [CrossRef]

23. Konsta-Gdoutos, M.S.; Aza, C.A. Self sensing carbon nanotube (CNT) and nanofiber (CNF) cementitious composites for real time damage assessment in smart structures. *Cem. Concr. Compos.* **2014**, *53*, 162–169. [CrossRef]
24. Han, B.H.; Guan, X.C.; Ou, J.P. Electrode design, measuring method and data acquisition system of carbon fiber cement paste piezoresistive sensors. *Sens. Actuators A* **2007**, *135*, 360–369. [CrossRef]
25. Azhari, F. Cement-based sensors for structural health monitoring. Ph.D. Thesis, University of British Columbia, Vancouver, BC, Canada, 2008.
26. Hou, T.C.; Lynch, J.P. Conductivity-based strain monitoring and damage characterization of fiber reinforced cementitious structural components. In *Smart Structures and Materials 2005: Sensors and Smart Structures Technologies for Civil, Mechanical, and Aerospace Systems*; International Society of Optics and Photonics: Bellingham, UK, 2005; pp. 419–430.
27. Ding, Y.N.; Han, Z.B.; Zhang, Y.L.; Aguiar, J.B. Concrete with triphasic conductive materials for self-monitoring of cracking development subjected to flexure. *Compos. Struct.* **2016**, *138*, 184–191. [CrossRef]
28. Han, B.G.; Zhang, L.Q.; Zhang, C.Y.; Wang, Y.Y.; Yu, X.; Ou, J.P. Reinforcement effect and mechanism of carbon fibers to mechanical and electrically conductive properties of cement-based materials. *Constr. Build. Mater.* **2016**, *125*, 479–489. [CrossRef]
29. Le, H.V.; Kim, D.J. Effect of matrix cracking on electrical resistivity of high performance fiber reinforced cementitious composites in tension. *Constr. Build. Mater.* **2017**, *156*, 750–760. [CrossRef]
30. Han, B.G.; Ou, J.P. Embedded piezoresistive cement-based stress/strain sensor. *Sens. Actuators A* **2007**, *138*, 294–298. [CrossRef]
31. Sun, S.W.; Han, B.G.; Shan, J.; Yu, X.; Wang, Y.L.; Li, H.Y.; Ou, J.P. Nano graphite platelets-enabled piezoresistive cementitious composites for structural health monitoring. *Constr. Build. Mater.* **2017**, *136*, 314–328. [CrossRef]
32. Wenner, F. A method for measuring earth resistivity. *J. Washington Acad. Sci.* **1915**, *5*, 561–563. [CrossRef]
33. Kim, M.K.; Kim, D.J. Electromechanical response of high-performance fiber-reinforced cementitious composites containing milled glass fibers under tension. *Materials* **2018**, *11*, 1115. [CrossRef] [PubMed]
34. Downey, A.; D'Alessandro, A.; Ubertini, F.; Laflamme, S.; Geiger, R. Biphasic DC measurement approach for enhanced measurement stability and multi-channel sampling of self-sensing multi-functional structural materials doped with carbon-based additives. *Smart. Mater. Struct.* **2017**, *26*, 065008. [CrossRef]
35. Suryanto, B.; McCarter, W.J.; Starrs, G.; Ludford-Jones, G.V. Electrochemical immittance spectroscopy applied to a hybrid PVA/steel fiber engineered cementitious composite. *Mater. Design.* **2016**, *105*, 179–189. [CrossRef]
36. Alexander, C.K.; Sadiku, M.N.O. Basic Concepts. In *Fundamentals of Electric Circuits*; McGraw-Hill Education: Columbus, OH, USA, 2013.

sensors

MDPI

Article

Facile Chemical Bath Synthesis of SnS Nanosheets and Their Ethanol Sensing Properties

Wei Shan [1,2], Zhengqian Fu [1], Mingsheng Ma [1], Zhifu Liu [1,2,*], Zhenggang Xue [3], Jiaqiang Xu [3], Faqiang Zhang [1] and Yongxiang Li [1,2,*]

1 CAS Key Lab of Inorganic Functional Materials and Devices, Shanghai Institute of Ceramics, Chinese Academy of Sciences, Shanghai 200050, China; shanwei@student.sic.ac.cn (W.S.); fmail600@mail.sic.ac.cn (Z.F.); mamingsheng@mail.sic.ac.cn (M.M.); zhangfq@mail.sic.ac.cn (F.Z.)

2 Center of Materials Science and Optoelectronics Engineering, University of Chinese Academy of Sciences, Beijing 100049, China

3 Department of Chemistry, Shanghai University, Shanghai 200444, China; xuezhenggang@163.com (Z.X.); xujiaqiang@shu.edu.cn (J.X.)

* Correspondence: liuzf@mail.sic.ac.cn (Z.L.); yxli@mail.sic.ac.cn (Y.L.); Tel.: +86-189-1622-1641 (Z.L.); +86-136-4172-3064 (Y.L.)

Received: 6 May 2019; Accepted: 27 May 2019; Published: 6 June 2019

Abstract: Tin(II) monosulfide (SnS) nanosheets were synthesized using $SnCl_4{\bullet}5H_2O$ and S powders as raw materials in the presence of H_2O via a facile chemical bath method. Orthorhombic phase SnS nanosheets with a thickness of ~100 nm and lateral dimensions of 2~10 μm were obtained by controlling the synthesis parameters. The formation of a SnO_2 intermediate is key to the valence reduction of Sn ions (from IV to II) and the formation of SnS. The gas sensors fabricated from SnS nanosheets exhibited an excellent response of 14.86 to 100 ppm ethanol vapor when operating at 160 °C, as well as fast response and recovery times of 23 s and 26 s, respectively. The sensors showed excellent selectivity for the detection of ethanol over acetone, methanol, and ammonia gases, which indicates the SnS nanosheets are promising for high-performance ethanol gas sensing applications.

Keywords: gas sensors; SnS nanosheets; chemical bath synthesis; oleylamine; ethanol sensing

1. Introduction

Ethanol (EtOH), a typical volatile organic compound (VOC), is a common raw material and has important applications in the chemical, biomedical, and food industries. However, it is volatile and flammable, and it can cause health problems such as headache and liver damage [1]. The detection of EtOH gas/vapor is therefore of particular importance. Nowadays, semiconductor gas sensors are most widely used due to their high sensitivity, ease of fabrication, simple structure, and low cost [2]. For example, SnO_2 [3], ZnO [4], In_2O_3 [5], and WO_3 [6] have been employed to fabricate ethanol gas sensors. However, most oxide semiconductor gas sensors tend to exhibit high operating temperatures [7–9] which increase the power consumption, accelerate the aging of gas-sensing materials, and reduce their service life. Recently, TiO_2 and Fe_2O_3 have been employed to make gas sensors which can work at room temperature [10–12]. Bhowmik et al. made an EtOH sensor device based on the p–n homojunction of TiO_2 nanostructures [10]; it offered a maximum response magnitude of ~57% (toward ethanol) at 100 ppm with appreciably fast response and recovery times of ~30 and ~16 s, while the nano-heterostructure increased the difficulty of device fabrication. Bindra et al. fabricated multi-layered TiO_2 nanotube-array-based highly sensitive room-temperature vapor sensors [11], but the response time was more than 82 s; Lin et al. made an EtOH gas sensor based on γ-Fe_2O_3 nanoparticles working at room temperature [12], but the response time was up to 533 s. In addition, most oxide semiconductor ethanol-gas-sensitive materials also have strong interference signals for other reducing VOCs, among which the most representative interference sources are acetone and methanol. At present, commercial EtOH sensors are composed of an array of gas-sensing materials with different selectivity, and the gas recognition and composition analysis are performed by pattern recognition technology [13,14], but these sensing arrays have complicated structures. So, it is still necessary to develop a gas-sensing material which has low operating temperature, high sensitivity, good selectivity, and a fast response time to EtOH gas.

Recently, flak-like sulfide materials have attracted great interest because of their excellent electrical, optical, and mechanical properties [15–17], as well as their planar crystal structure, large surface-to-volume ratio, and low electronic noise [18,19]. These properties make sulfide nanosheets very suitable for gas-sensing materials; this has not received the deserved attention. In most cases, sulfide-nanosheet-based gas sensors have a lower operating temperature than do metal oxide semiconductor materials [19–21]. This is mainly due to the S atoms exposed on the surface of sulfide nanosheets that have less electronegativity and more easily absorb oxygen in the air. At the same time, the narrower band gap of the sulfide material (<2.5 eV), compared with that of oxide semiconductor materials (~3.0 eV), can reduce the power consumption of gas sensors. Additionally, previous research based on first-principles calculations and experiments has shown that physisorption-based charge transfer functions more easily between the surface of flaky sulfide gas-sensitive materials and target gas molecules at lower temperatures, which may result in better gas sensing selectivity for a particular target gas [20,22,23].

As a new 2D material with a black phosphorene structure, SnS has an energy band gap of about 1.3 eV [24,25], which is suitable for a gas-sensing material. Recently, Rana et al. fabricated a gas sensor based on SnS nanoparticles [26] which had high sensitivity and a fast response time to EtOH gas. This confirmed that SnS is a very promising ethanol gas-sensing material. Three kinds of methods have been used to synthesize SnS nanomaterial: top-down exfoliation from bulk materials, chemical vapor deposition (CVD), and solution synthesis [27–32]. However, the nanosheets obtained by exfoliation methods from bulk materials are not controllable in terms of their uniformity, size, and thickness, which is not suitable for large-scale production. Although O'Brien et al. prepared SnS thin films via CVD [29], the complex precursors and strict synthesis conditions limited the application of this method. Herron et al. prepared SnS nanosheets by a solution method, but the particles were small and thick [30]. Deng et al. synthesized ultrathin single crystalline SnS in oleylamine (OAm) but with the addition of 1-dodecanethiol (DDT), which is a highly toxic pesticide [31]. Cui et al. prepared large-sized SnS thin crystals by solvothermal methods; however, the reaction pressure was up to 200 MPa, which is hard to achieve and dangerous in operation [32]. Therefore, there is still a significant need to explore a simple, efficient, and environmentally friendly process to grow large-sized SnS nanosheets.

In the present work, we report a facile chemical bath method to synthesize large-sized SnS nanosheets from cheap and environmentally benign chemicals. The gas sensing results of the synthesized SnS nanosheets show a high response, great selectivity, and fast response and recovery times towards EtOH gas.

2. Materials and Methods

2.1. SnS Nanosheet Synthesis and Characterization

Tin (IV) chloride ($SnCl_4{\bullet}5H_2O$, >99.9%, Aladdin) and sulfur powder (5N, Aladdin) were used as the tin (Sn) and sulfide (S) precursors, respectively; oleylamine (OAm, >90.0%, Aladdin) was used as a surfactant; and octadecene (Ode, >90.0%, Aladdin) was the solvent. All the chemical reagents were used as received without further purification. Figure 1 shows a diagram of the synthesis process. A simplified Schlenk line was used to protect the reaction from oxygen/air, and the whole synthesis process should be performed under a persistent flow of high-purity N_2 gas with constant stirring.

In a typical experiment, $SnCl_4{\bullet}5H_2O$ (1 mmol) was dissolved into distilled water (10 mL), then the above solution and Ode (20 mL) were added into a 50 mL three-neck flask. This mixed solution was degassed at 130 °C for 1 h to fully hydrolyze $SnCl_4 \bullet 5H_2O$ and evaporate excess water. Subsequently, the solution was heated to 280 °C within 15 min. The S-OAm solution (1 mmol of sulfur powder dispersed into 10 mL of OAm) was injected into the reaction system, and the reaction was maintained at 280 °C for 30 min. After cooling the solution to room temperature naturally, a gray-black product was collected and separated from the solution by centrifugation. The product was further washed three times using an ethanol and cyclohexane mixed solvent (1:1 volume ratio) and finally dispersed in cyclohexane. The morphology of the products was examined by scanning electron microscopy (SEM, JEOL S4800) equipped with an energy-dispersive X-ray (EDX) spectroscope and a transmission electron microscope (TEM, JEM-2100F, JEOL). Powder X-ray diffraction (XRD) patterns were taken on an X'Pert Pro MPD (Philips PAN analytical) with Cu Kα radiation at 45 kV and 40 mA. The XRD data were collected with the θ-2θ scanning scope starting from 2θ = 5° to 80° with a step of 6° per second by using Cu Kα radiation (1.5418 Å). The composition of the reaction liquid was characterized by Fourier transform infrared spectrometry (FT-IR, 8400 Shimadzu).

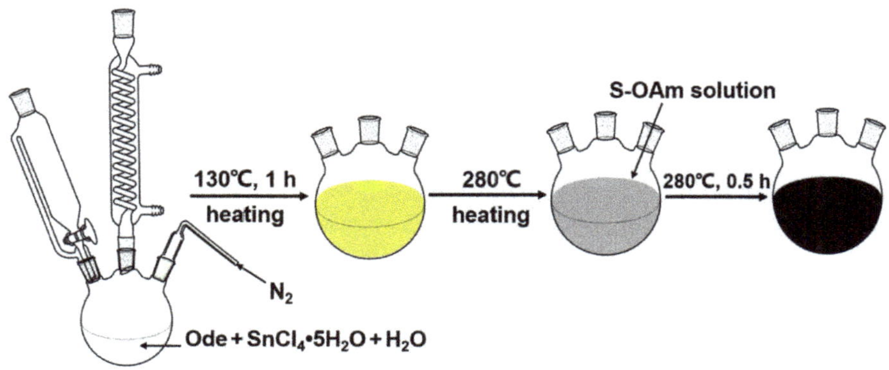

Figure 1. Diagram of the chemical bath synthesis process of SnS nanosheets.

2.2. The Gas Sensing Test

To make the gas sensing device, 0.1 g of as-prepared SnS nanosheets and 5 mL of cyclohexane were added into a centrifuge tube with 5 mL capacity. This mixture was ultrasonicated for 10 min to form a concentrated suspension. An alumina ceramic tube (Figure 2) with two ring-shaped gold electrodes and two platinum wires connected to each electrode were put into the suspension such that the surface of the tube was coated with SnS nanosheets and then dried in ambient air naturally. Repeating the above processes five times, a gas sensing test tube whose surface was completely and evenly covered by SnS nanosheets was prepared. Then, a Ni–Cr alloy filament was put into the ceramic tube as a micro heater. The operating temperature of the gas sensor is controlled by the voltage across the heating wire. The final gas sensing test unit was assembled by soldering the ceramic tube with SnS nanosheet sensing material and electrodes onto a connector base with six probes.

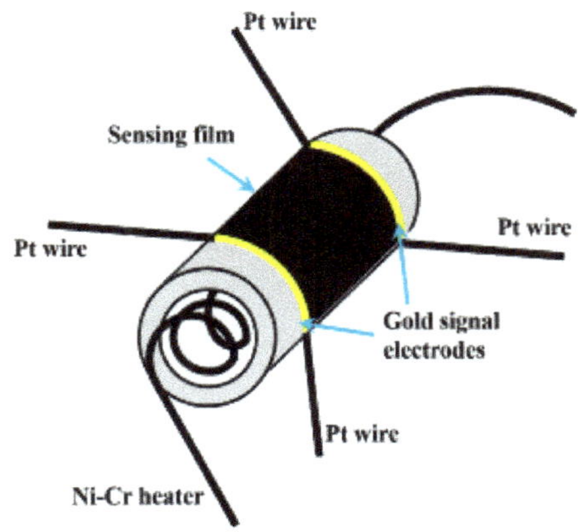

Figure 2. Schematic diagram of the device for gas sensing performance testing.

The gas sensing tests were carried out using a commercial HW-30A measurement system (Han Wei Electronics Co., Zhengzhou, China) using ambient air as the dilute and reference gas. For details on this test system, please see the Supplementary Materials. A hygrometer showed a humidity of 60% at room temperature (25 °C). The response time (t_{res}) or recovery time (t_{rec}) is expressed as the time taken for the sensor output to reach 90% of its saturation after the test gas is applied or deactivated in a function step. The response S was defined as

$$S = R_g/R_a, \quad (1)$$

where R_a and R_g are the steady-state resistances in air and the testing gas, respectively.

3. Results and Discussion

3.1. Microstructural Characterizations of the SnS Nanosheets

Figure 3a shows the crystal structure diagram of SnS. Figure 3b is the XRD pattern of as-prepared typical SnS nanosheets, which share the same structure as orthorhombic phase SnS (JCPDS No. 39-0354) with a space group of Pbnm(62). All sharp peaks match the JCPDS card, and no impurities were identified. In this crystal structure, S atoms and Sn atoms in the same two-atom-thick layer are alternately arranged in a zigzag (Figure 3a), and two adjacent layers are held together by van der Waals forces [33]. Each unit cell contains two of these two-atom-thick layers along the b axis. The primary diffraction peak at the 2θ value of 31.97°, ascribed to the (040) plane, is consistent with the SEM and TEM images in Figure 4. The SEM images (Figure 4a,b) display the morphology of typical SnS nanosheets. Most as-prepared SnS particles are sheet-like with lateral sizes of 5~20 μm and a thickness of about 100 nm. Figure 4c shows a TEM image of a monodisperse SnS nanosheet, and Figure 4d is a high-resolution TEM (HRTEM) picture of the square area in Figure 4c. The corresponding selected area electron diffraction (SAED) pattern reveals the single-crystalline structure of orthorhombic phase SnS nanosheets with the extending direction parallel to the (040) plane. Figure 4e shows an HRTEM image of the region circled by the frame in Figure 4d, and the inset is a profile intensity image along the line, in which a lattice fringe spacing of 0.293 nm is identified. It should correspond to the ($\bar{1}01$) or (101) lattice planes of orthorhombic phase SnS. EDX analysis was performed and strong signals from Sn and S were detected, as shown in Figure 4f. The peaks related to Cu came from the copper mesh substrate used in the TEM analysis. The calculated composition ratio was 1.03(Sn):0.97(S), with a slight sulfur deficiency compared to the stoichiometric SnS. It is important that the deficient S atoms make the as-prepared SnS nanosheets exhibit n-type semiconducting characteristics. The high aspect ratio and large area of the high-quality (040) surface endue it with promising applications like catalysis and gas sensing.

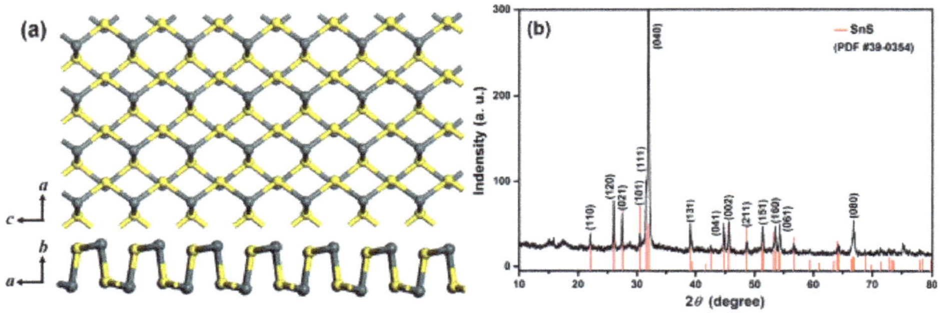

Figure 3. (**a**) Crystal structure diagram of orthorhombic phase SnS. (**b**) XRD pattern of typical SnS nanosheets.

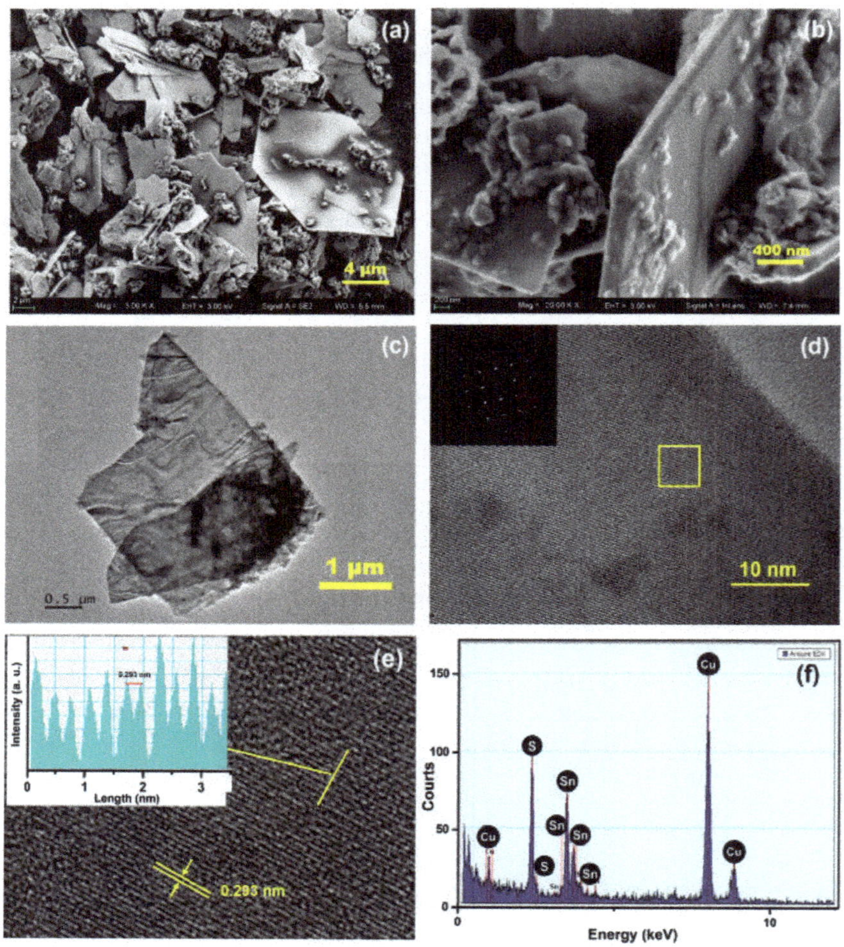

Figure 4. (**a**) and (**b**) SEM images of SnS nanosheets. (**c**) TEM image of SnS nanosheets. (**d**) and (**e**) HRTEM images of SnS nanosheets. (**f**) The EDX spectrum of SnS nanosheets.

3.2. Synthesis Process and Mechanism

To understand the reaction process, especially the valence conversion of Sn ions from +4 to +2, we conducted a series of characterizations to different time-dependent intermediate products. Firstly, a series of samples were prepared with different volumes of water (2 mL, 4 mL, 6 mL, 8 mL, and typical 10 mL) while the other conditions remained unchanged. Figure 5 shows the XRD patterns of these samples. When there is no water addition during the reaction, the main products are hexagonal crystal SnS_2 flakes. However, the main product changed to orthorhombic crystal SnS when even a small amount of water was introduced into the reaction system. With an increasing amount of water, the yield of SnS in the product increases. Pure SnS nanosheets were obtained when 10 mL water was added in the reaction. This demonstrates that the added water is not only a solvent but also participates in the chemical reaction.

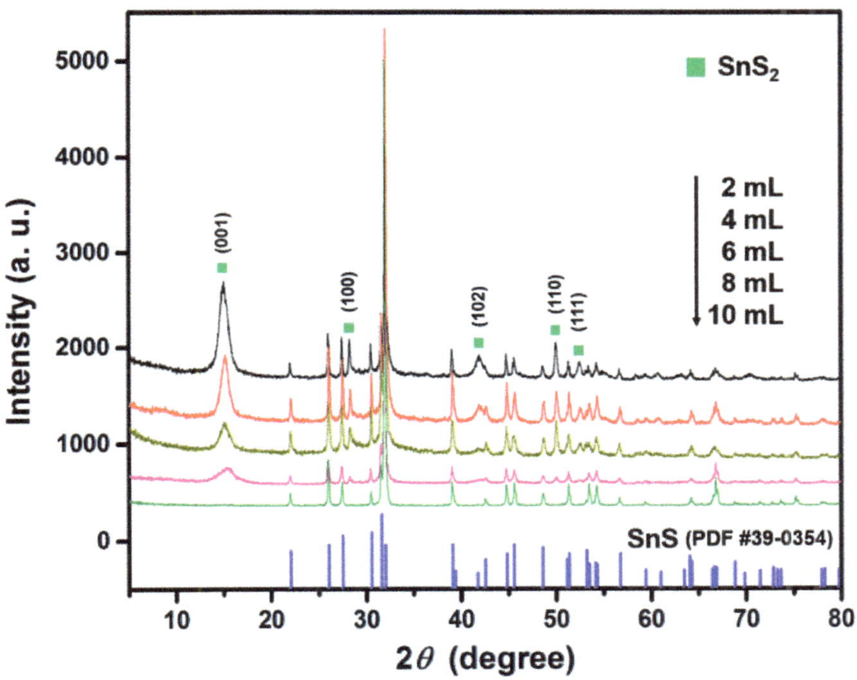

Figure 5. XRD patterns of samples synthesized with different volumes of water.

The intermediate product was further analyzed in order to understand the function of water in the reaction process. Figure 6 shows an SEM image of a sample obtained by heating the reactants at 130 °C for 1 h and then by centrifugation. It is apparently not a crystallized substance. The EDS analysis results indicate that it contains Sn, O, and Cl elements with an atom ratio of about 1(Sn):3(O):1(Cl). In addition, we passed the effluent gas through a flask with deionized water. After the reaction, the testing water became acidic, which indicates the formation of hydrochloric acid (HCl). It is possible that $SnCl_4$ was hydrolyzed during the heat preservation process following the chemical reaction

$$SnCl_4 + H_2O \overset{\Delta}{\Rightarrow} Sn(OH)_3Cl + HCl \uparrow . \quad (2)$$

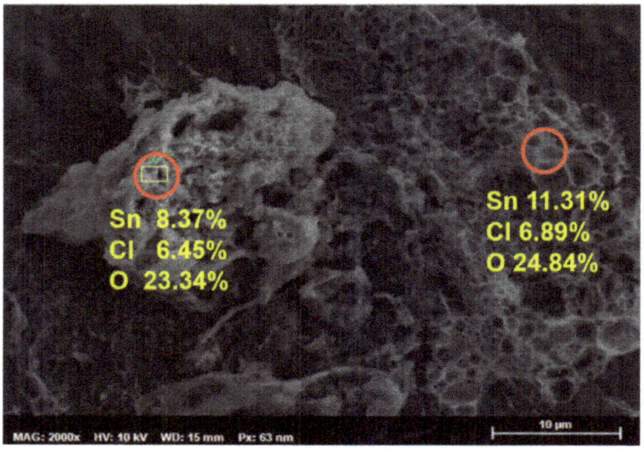

Figure 6. SEM image of $SnCl_4$ hydrolyzates and the element composition of the circled area measured by EDX.

Moreover, the reaction was stopped when the reaction temperature had just reached 280 °C and the S-OAm solution had not been injected into the reaction flask. The reaction mixture was centrifuged and washed with EtOH. Finally, reddish-black powders were obtained. The XRD pattern shown in Figure 7 indicates that the powder is a tetragonal crystal of SnO_2 (JCPDS No. 41-1445) with a low degree of crystallinity. It is possible that the intermediate compound $Sn(OH)_3Cl$ was decomposed to form SnO_2 at an elevated temperature (280 °C). After injection of the S-OAm solution into the reaction system, both reactions (4) and (5) may take place [34]. The SnS nanosheets can be formed following reaction (6).

$$Sn(OH)_3Cl \rightarrow SnO_2 + HCl + H_2O \tag{3}$$

$$R\text{-}CH_2\text{-}NH_2 + SnO_2 \overset{\Delta}{\Rightarrow} R\text{-}CO\text{-}NH_2 + SnO + H_2O \tag{4}$$

$$R\text{-}CH_2\text{-}NH_2 + S_x \overset{\Delta}{\Rightarrow} R\text{-}CS\text{-}NH_2 + H_2S \tag{5}$$

$$SnO + H_2S \overset{\Delta}{\Rightarrow} SnS + H_2O \tag{6}$$

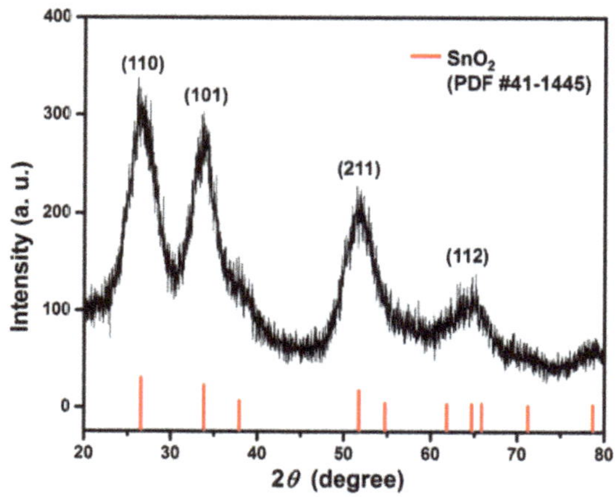

Figure 7. XRD pattern of the intermediate product before S-OAm reactant was added.

In order to verify this assumption, the suspension was allowed to stand for 6 h after completion of the reaction, and then the supernatant liquid was taken for FT-IR analysis. The FTIR spectrum is shown in Figure 8. The obvious peaks at 1560 and 1647 cm^{-1} are attributed to the R-CO–NH_2 stretch of amide, which confirms the formation of amide during the chemical bath reaction [35]. As discussed above, it could be concluded that reactions (2)–(4) are the key to the valence conversion of the Sn ion from +4 to +2.

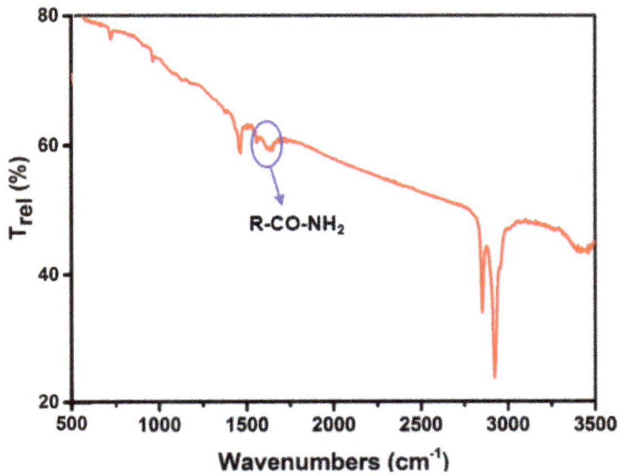

Figure 8. FI-IR spectrum of the final reaction solution.

3.3. Gas Sensing Properties

A large number of studies have demonstrated that the operating temperature of a gas sensor is the most influential factor which affects the gas response of semiconductor sensors [36]. It is known that sulfides are unstable in air at high temperatures, so we examined the temperature stability of the as-prepared SnS nanosheets by heating the samples to different temperatures. With the help of XRD characterization, we could clearly observe the transition from orthorhombic phase SnS to SnO_x after heating the nanosheets to 180 °C in air for 10 days, but this phenomenon did not appear with the same process at 160 °C, as shown in Figure 9a,b. Therefore, we tested the gas sensing response of SnS nanosheets to 100 ppm EtOH at operating temperatures from room temperature to 160 °C, and the results are shown in Figure 9c. At room temperature, there is no acceptable signal because the resistance of the sensing device is too big for the measurement range of the instrument. With increasing operating temperature, the response is enhanced, suggesting that increased temperature facilitates the adsorption of EtOH gas molecules onto the surface of SnS nanosheets and promotes the charge transfer [17]. This positive correlation between response *S* and operating temperature is valid until the temperature of 160 °C, at which the response *S* is 14.86. At present, some of the conversional metal oxide semiconductor (ZnO, WO_3, etc.) gas sensors work at over 250 °C [1,37]. The low operating temperature of the SnS nanosheet gas sensor may be due to the following reasons. On one hand, the electronegativity of the S atom is lower than that of the O atom. Thus, compared with the surface of metal oxides, oxygen molecules are more easily adsorbed on the surface of tin sulfides and then capture electrons from the conduction band to form oxygen ions; this makes sensing signals based on chemical adsorption easy to produce. On the other hand, the physisorption-based charge transfer also contributes significantly to the gas sensing properties of flaky sulfide materials, which can work at lower temperatures [18,38].

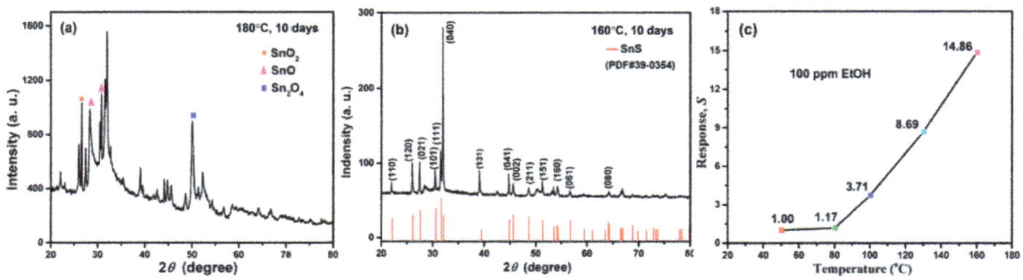

Figure 9. (**a**) XRD pattern of SnS samples which underwent heat treatment at 180 °C for 10 days in air. (**b**) XRD pattern of SnS samples which underwent heat treatment at 160 °C for 10 days in air. (**c**) The response *S* of sensors made of SnS nanosheets in the presence of 100 ppm of EtOH gas operating at different temperatures.

Figure 10a shows the dynamic response–recovery curves of the sensors to EtOH gas with concentrations ranging from 1 ppm to 100 ppm at 160 °C. Resistances of 35.7, 11.9, 7.9, 5.6, and 3.7 MΩ were recorded at EtOH gas concentrations of 1, 10, 20, 50, and 100 ppm, respectively. With decreasing EtOH concentration, the response signal is getting weaker and weaker. This is due to fewer ethanol molecules being adsorbed on the surface of the sensing material, and the resistance of the SnS nanosheets changes less. When the concentration of EtOH gas decreases to 1 ppm, the gas sensor still shows an obvious response *S* of 1.54, revealing that the resistance of the SnS nanosheets is very sensitive to EtOH vapor. The relationship between the response *S* and concentrations of NO_2 is shown in the Supplementary Materials.

In order to study their reversibility characteristics, the sensing devices were exposed to 100 ppm of EtOH and air alternately for five or more cycles. It can be seen in Figure 10b that the resistance of the sensor in 100 ppm EtOH gas was steady and could return to the no-load level after each desorption. This result confirms that there is a minuscule amount of residual gas molecules during desorption process, and the SnS nanosheets gas sensors have a good performance in reversible cycle tests. The response and recovery times of the sensors to 100 ppm EtOH at 160 °C are shown in Figure 10c. As can be seen, the response and recovery times are 23 s and 26 s, respectively. These results are much better than those of most EtOH gas-sensing materials reported previously, such as Bi_2S_3 nanowires [39], CuxO films [40], WO_3 microflowers [1], and V_2O_5 nanobelts [41].

Figure 11 shows the response *S* of SnS-nanosheet-based sensors to 100 ppm of different gases at 160 °C. Compared to EtOH, other gases (including acetone and methanol gases, which are the most common sources of signal interference) have little response with the SnS-nanosheet-based gas sensor. In this work, the response *S* to EtOH is more than 10 times larger than those to the other gases. This indicates the SnS-nanosheet-based sensors exhibit a superior selectivity to EtOH gas. Ethanol gas is considered a representative of the reducing organic gases. According to previous studies, for volatile organic compounds (VOCs), the optimal operating temperatures depend on their molecular orbital levels [40]. When the lowest unoccupied molecular orbital (LUMO) level of the gas is low, the energy required for the gas sensing reaction is reduced. Since the LUMO level of ethanol (0.12572 eV) is lower than those of methanol (0.19728 eV) and acetone (0.20525 eV), electrons in the ethanol molecule are more prone to transfer than those in methanol and acetone molecules at low temperature [42,43]. In addition, physisorption-based charge transfer may play an important role in the gas sensing process. In this case, the more the electron cloud density shifts, the higher the molecule–surface binding energy becomes; this is reflected in the macro phenomenon in which the material resistance changes more and the gas sensing signal is stronger. Therefore, we speculate that the SnS nanosheets we prepared have higher binding energy to ethanol molecules than other tested organic molecules, especially methanol and acetone. This requires further experimentation or calculation to verify. As shown above, the SnS-nanosheet-based gas sensor in this study showed excellent selectivity to ethanol molecules.

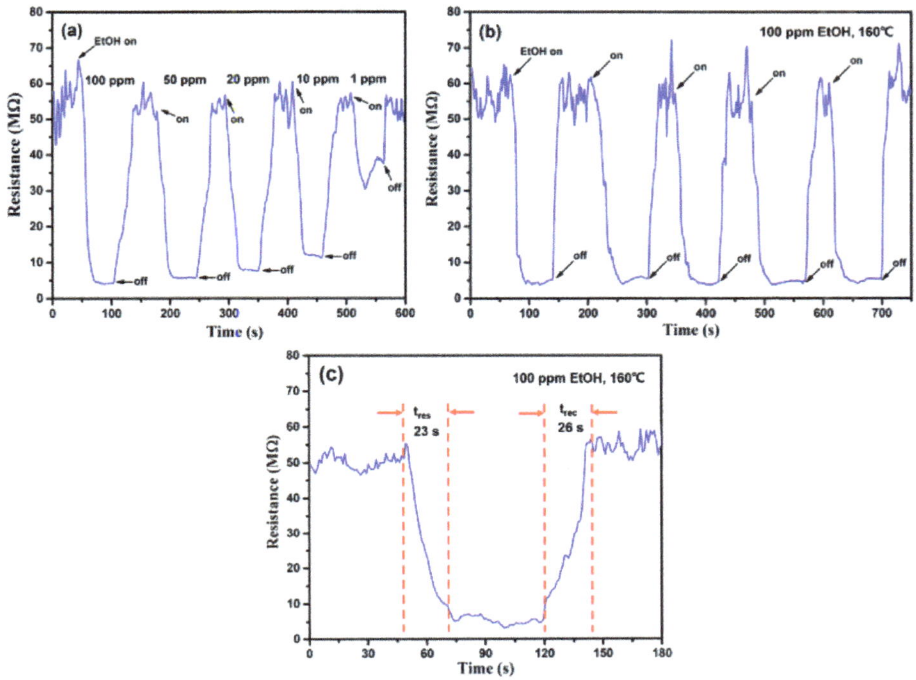

Figure 10. Gas sensing performance of SnS nanosheets: (**a**) Dynamic sensing performance of SnS nanosheets toward EtOH gas at concentrations ranging from 1 ppm to 100 ppm at 160 °C. (**b**) Five reversible test cycles of SnS nanosheets toward 100 ppm EtOH at 160 °C. (**c**) A single cycle test of SnS nanosheets after exposure to 100 ppm EtOH at 160 °C.

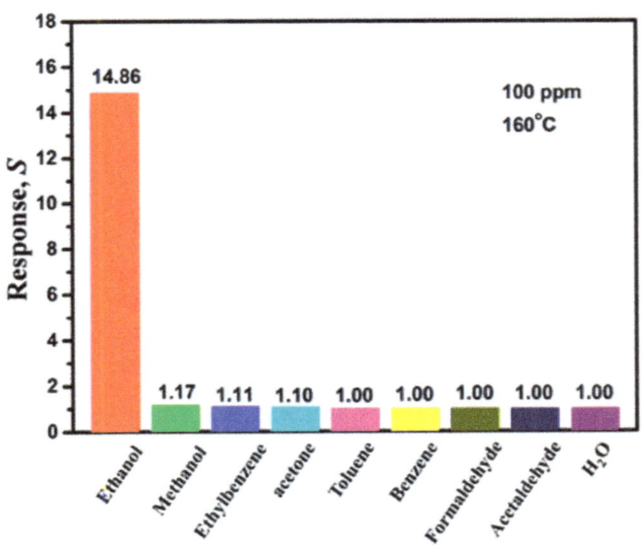

Figure 11. The response S towards various test gases at 100 ppm of SnS nanosheets at an operating temperature of 160 °C.

4. Conclusions

In this work, SnS nanosheets were prepared using a simple solution reaction method. Orthorhombic phase SnS nanosheets with a thickness of ~100 nm and lateral dimensions of about 2~10 μm were obtained by controlling the synthesis parameters. The formation of a SnO_2 intermediate is key to the valence reduction of Sn (from IV to II) and the formation of SnS nanosheets. The gas sensing properties of the SnS nanosheets were investigated at different operating temperatures. The SnS-nanosheet-based gas sensors exhibited excellent properties in sensing EtOH gas with a maximum response *S* of 14.86 at 160 °C, and a response time of 23 s and recovery time of 26 s. The response factor to EtOH is more than 10 times those to other VOCs. This study demonstrates the potential of SnS nanosheets for EtOH gas sensing applications.

Supplementary Materials: The following are available online at http://www.mdpi.com/1424-8220/19/11/2581/s1, Figure S1: Photo of (a) the test circuit board and (b) HW-30A gas-sensing test system, Figure S2. Relationship between response S and concentrations of NO2 at 110 °C.

Author Contributions: Conceptualization, Y.L. and Z.L.; data curation, W.S. and Z.L.; formal analysis, W.S., Z.F., F.Z., Z.X. and J.X.; funding acquisition, M.M., F.Z. and Z.L.; investigation, W.S.; methodology, W.S. and Y.L.; project administration, Z.L. and Y.L.; resources, W.S.; software, W.S.; supervision, Z.L. and Y.L.; validation, W.S.; writing—original draft, W.S.; writing—review and editing, Z.L. and Y.L.

Funding: The National Natural Science Foundation of China (NSFC, Nos. 51502325), and M.S. Ma acknowledges the Youth Innovation Promotion Association of CAS.

Acknowledgments: In this section you can acknowledge any support given which is not covered by the author contribution or funding sections. This may include administrative and technical support, or donations in kind (e.g., materials used for experiments).

Conflicts of Interest: The authors declare no conflict of interest.

References

1. Mehta, S.S.; Tamboli, M.S.; Mulla, I.S.; Suryavanshi, S.S. Facile hydrothermal synthesis of nanobricks assembled WO_3 microflowers and their ethanol sensing properties. *Mater. Lett.* **2017**, *207*, 80–84. [CrossRef]
2. Saboor, F.H.; Khodadadi, A.A.; Mortazavi, Y.; Asgari, M. Microemulsion synthesized Silica/ZnO stable core/shell sensors highly selective to ethanol with minimum sensitivity to humidity. *Sens. Actuators B Chem.* **2017**, *238*, 1070–1083. [CrossRef]
3. Chen, Y.J.; Xue, X.Y.; Wang, Y.G.; Wang, T.H. Synthesis and ethanol sensing characteristics of single crystalline SnO_2 nanorods. *Appl. Phys. Lett.* **2005**, *87*, 233503. [CrossRef]
4. Wang, L.W.; Kang, Y.F.; Liu, X.H.; Zhang, S.M.; Huang, W.P.; Wang, S.R. ZnO nanorod gas sensor for ethanol detection. *Sens. Actuators B Chem.* **2012**, *162*, 237–243. [CrossRef]
5. Lai, H.Y.; Chen, T.H.; Chen, C.H. Architecture controlled synthesis of flower-like In_2O_3 nanobundles with significantly enhanced ultraviolet scattering and ethanol sensing. *Cryst. Eng. Commun.* **2012**, *14*, 5589–5595. [CrossRef]
6. Li, X.L.; Lou, T.J.; Sun, X.M.; Li, Y.D. Highly sensitive WO_3 hollow-sphere gas sensors. *Inorg. Chem.* **2004**, *43*, 5442–5449. [CrossRef] [PubMed]
7. Baruwati, B.; Kumar, D.K.; Manorama, S.V. Hydrothermal synthesis of highly crystalline ZnO nanoparticles: A competitive sensor for LPG and EtOH. *Sens. Actuators B Chem.* **2006**, *119*, 676–682. [CrossRef]
8. Bagheri, M.; Hamedania, N.F.; Mahjouba, A.R.; Khodadadi, A.A.; Mortazavi, Y. Highly sensitive and selective ethanol sensor based on Sm_2O_3-loaded flower-like ZnO nanostructure. *Sens. Actuators B Chem.* **2014**, *191*, 283–290. [CrossRef]
9. Song, P.; Han, D.; Zhang, H.H.; Li, J.; Yang, Z.X.; Wang, Q. Hydrothermal synthesis of porous In_2O_3 nanospheres with superior ethanol sensing properties. *Sens. Actuators B Chem.* **2014**, *196*, 434–439. [CrossRef]
10. Bhowmik, B.; Dutta, K.; Bhattacharyya, P. An efficient room temperature ethanol sensor device based on p-n homojunction of TiO_2 nanostructures. *IEEE Trans. Electron Devices* **2019**, *66*, 1063–1068. [CrossRef]
11. Bindra, P.; Hazra, A. Multi-layered TiO_2 nanotubes array based highly sensitive room temperature vapor sensors. *IEEE Trans. Nanotechnol.* **2019**, *18*, 13–20. [CrossRef]
12. Hou, L.; Zhang, C.M.; Ma, P.; Li, L.; Zhu, K.K.; Kang, X.F.; Chen, W. Ethanol gas sensor based on γ-Fe_2O_3 nanoparticles working at room temperature with high sensitivity. *Chin. J. Anal. Chem.* **2018**, *46*, 1854–1862. [CrossRef]

13. Carey, W.P.; Beebe, K.R.; Sanchez, E.; Geladi, E.P.; Kowalski, B.R. Chemometric analysis of multisensor arrays. *Sens. Actuators B Chem.* **1984**, *9*, 225–243. [CrossRef]
14. Pineau, N.J.; Kompalla, J.F.; Guntner, A.T.; Pratsinis, S.E. Orthogonal gas sensor arrays by chemoresistive material design. *Microchim. Acta* **2018**, *185*, 563. [CrossRef] [PubMed]
15. Novoselov, K.S.; Geim, A.K.; Morozov, S.V.; Jiang, D.; Zhang, Y.; Dubonos, S.V.; Grigorieva, I.V.; Firsov, A.A. Electric field effect in atomically thin carbon films. *Science* **2004**, *306*, 666–669. [CrossRef] [PubMed]
16. Huang, X.; Yin, Z.; Wu, S.; Qi, X.; He, Q.; Zhang, Q.; Yan, Q.; Boey, F.; Zhang, H. Graphene-based materials: Synthesis, characterization, properties, and applications. *Small* **2001**, *7*, 1876–1902. [CrossRef]
17. Novoselov, K.S.; Jiang, D.; Schedin, F.; Booth, T.J.; Khotkevich, V.V.; Morozov, S.V.; Geim, A.K. Two-dimensional atomic crystals. *Proc. Natl. Acad. Sci. USA* **2005**, *102*, 10451–10453. [CrossRef] [PubMed]
18. Seo, J.W.; Jang, J.T.; Park, S.W.; Kim, C.; Park, B.; Cheon, J. Two-dimensional SnS_2 nanoplates with extraordinary high discharge capacity for lithium ion batteries. *Adv. Mater.* **2008**, *20*, 4269–4273. [CrossRef]
19. Kim, Y.H.; Phan, D.T.; Ahn, S.; Namb, K.H.; Park, C.M.; Jeon, K.J. Two-dimensional SnS_2 materials as high-performance NO_2 sensors with fast response and high sensitivity. *Sens. Actuators B Chem.* **2018**, *255*, 616–621. [CrossRef]
20. Ou, J.Z.; Ge, W.Y.; Carey, B.; Daeneke, T.; Rotbart, A.; Shan, W.; Wang, Y.C.; Fu, Z.Q.; Chrimes, A.F.; Wlodarski, W.; et al. Physisorption-based charge transfer in two-dimensional SnS_2 for selective and reversible NO_2 gas sensing. *Am. Chem. Soc. Nano* **2015**, *9*, 10313–10323.
21. Perkins, F.K.; Friedman, A.L.; Cobas, E.; Campbell, P.M.; Jernigan, G.G.; Jonker, B.T. Chemical Vapor Sensing with Monolayer MoS_2. *Nano Lett.* **2013**, *13*, 668–673. [CrossRef] [PubMed]
22. Ma, S.; Yuan, D.; Wang, Y.; Jiao, Z. Monolayer GeS as a potential candidate for NO_2 gas sensors and capturers. *J. Mater. Chem. C* **2018**, *6*, 8082–8091. [CrossRef]
23. Rana, C.; Bera, S.R.; Saha, S. Growth of SnS nanoparticles and its ability as ethanol gas sensor. *J. Mater. Sci. Mater. Electron.* **2019**, *30*, 2016–2029. [CrossRef]
24. Barzegar, M.; Tudu, B. Two-dimensional materials for gas sensors: From first discovery to future possibilities. *Surf. Innov.* **2018**, *6*, 1–79. [CrossRef]
25. Vidal, J.; Lany, S.; D'Avezac, M.; Zunger, A.; Zakutayev, A.; Francis, J.; Tate, J. Band-structure, optical properties, and defect physics of the photovoltaic semiconductor SnS. *Appl. Phys. Lett.* **2012**, *100*, 032104. [CrossRef]
26. Sinsermsuksakul, P.; Heo, J.; Noh, W.; Hock, A.S.; Gordon, R.G. Atomic layer deposition of tin monosulfide thin films. *Adv. Energy Mater.* **2011**, *1*, 1116–1125. [CrossRef]
27. Wangperawong, A.; Herron, S.M.; Runser, R.R.; Hagglund, C.; Tanskanen, J.T.; Lee, H.B.R.; Clemens, B.M.; Bent, S.F. Vapor transport deposition and epitaxy of orthorhombic SnS on glass and NaCl substrates. *Appl. Phys. Lett.* **2013**, *103*, 052105. [CrossRef]
28. Wang, Q.H.; Kalantar-Zadeh, K.; Kis, A.; Coleman, J.N.; Strano, M.S. Electronics and optoelectronics of two-dimensional transition metal dichalcogenides. *Nat. Nanotechnol.* **2012**, *7*, 699–712. [CrossRef]
29. Ramasamy, K.; Kuznetsov, V.L.; Gopal, K.; Malik, M.A.; Raftery, J.; Edwards, P.P.; O'Brien, P. Organotin dithiocarbamates: Single-source precursors for tin sulfide thin films by aerosol assisted chemical vapor deposition (AACVD). *Chem. Mater.* **2013**, *25*, 266–276. [CrossRef]
30. Herron, S.M.; Tanskanen, J.T.; Roelofs, K.E.; Bent, S.F. Highly textured tin(II) sulfide thin films formed from sheetlike nanocrystal inks. *Chem. Mater.* **2014**, *26*, 7106–7113. [CrossRef]
31. Deng, Z.T.; Cao, D.; He, J.; Lin, S.; Lindsay, S.M.; Liu, Y. Solution synthesis of ultrathin single crystalline SnS nanoribbons for photodetectors via phase transition and surface processing. *Am. Chem. Soc. Nano* **2012**, *6*, 6197–6207. [CrossRef] [PubMed]
32. Wang, J.; Lian, G.; Xu, Z.H.; Fu, C.; Lin, Z.J.; Li, L.Y.; Wang, Q.L.; Cui, D.L.; Wong, C.P. Growth of large-size SnS thin crystals driven by oriented attachment and applications to gas sensors and photodetectors. *Am. Chem. Soc. Appl. Mater. Interfaces* **2016**, *8*, 9545–9551. [CrossRef] [PubMed]
33. Zhang, Y.J.; Lu, J.; Shen, S.L.; Xu, H.R.; Wang, Q.B. Ultralarge single crystal SnS rectangular nanosheets. *Chem. Commun.* **2011**, *47*, 5226–5228.
34. Thomson, J.W.; Nagashima, K.; Macdonald, P.M.; Ozin, G.A. From sulfur-amine solutions to metal sulfide nanocrystals: Peering into the oleylamine-sulfur black box. *J. Am. Chem. Soc.* **2011**, *133*, 5036–5041. [CrossRef]
35. Du, Y.P.; Yin, Z.Y.; Rui, X.H.; Zeng, Z.Y.; Wu, X.J.; Liu, J.Q.; Zhu, Y.Y.; Zhu, J.X.; Huang, X.; Yan, Q.Y.; et al. A facile, relative green, and inexpensive synthetic approach toward large-scale production of SnS_2 nanoplates for high-performance lithium-ion batteries. *Nanoscale* **2013**, *5*, 1456–1459. [CrossRef] [PubMed]

36. Shi, J.J.; Cheng, Z.X.; Gao, L.P.; Zhang, Y.; Xu, J.Q.; Zhao, H.B. Facile synthesis of reduced graphene oxide/hexagonal WO_3 nanosheets composites with enhanced H_2S sensing properties. *Sens. Actuators B Chem.* **2016**, *230*, 736–745. [CrossRef]
37. Jing, Z.H.; Zhan, J.H. Fabrication and gas-sensing properties of porous ZnO nanoplates. *Adv. Mater.* **2008**, *20*, 4547–4551. [CrossRef]
38. Sing, K.S.W.; Everett, D.H.; Haul, R.A.W.; Moscou, L.; Pierotti, R.A.; Rouquerol, J.; Siemieniewska, T. Reporting physisorption data for gas/solid systems with special reference to the determination of surface area and porosity. *Pure App. Chem.* **1985**, *57*, 603–619. [CrossRef]
39. Yang, X.Y.; Tian, S.S.; Li, R.; Wang, W.; Zhou, S.M. Use of single-crystalline Bi_2S_3 nanowires as room temperature ethanol sensor synthesized by hydrothermal approach. *Sens. Actuators B Chem.* **2017**, *241*, 210–216. [CrossRef]
40. Zoolfakar, A.S.; Ahmad, M.Z.; Rani, R.A.; Ou, J.Z.; Balendhran, S.; Zhuiykov, S.; Latham, K.; Wlodarskia, W.; Kalantar-zadeh, K. Nanostructured copper oxides as ethanol vapour sensors. *Sens. Actuators B Chem.* **2013**, *185*, 620–627. [CrossRef]
41. Liu, J.F.; Wang, X.; Peng, Q.; Li, Y.D. Vanadium pentoxide nanobelts: Highly selective and stable ethanol sensor materials. *Adv. Mater.* **2005**, *17*, 764–767. [CrossRef]
42. Wen, Z.; Tian-Mo, L. Gas-sensing properties of SnO_2–TiO_2-based sensor for volatile organic compound gas and its sensing mechanism. *Physical. B* **2010**, *405*, 1345–1348. [CrossRef]
43. Wang, L.; Deng, J.; Fei, T. Template-free synthesized hollow NiO–SnO_2 nanospheres with high gas-sensing performance. *Sens. Actuators B Chem.* **2012**, *164*, 90–95. [CrossRef]

sensors

Article

Relaxation-Related Piezoelectric and Dielectric Behavior of Bi(Mg,Ti)O_3–PbTiO_3 Ceramic

Min Young Park, Jae-Hoon Ji and Jung-Hyuk Koh *

School of Electrical and Electronics Engineering, Chung-Ang University, Heukseok-Ro 84, Seoul 06974, Korea; pmy730@naver.com (M.Y.P.); hoon2441@naver.com (J.-H.J.)

* Correspondence: jhkoh@cau.ac.kr; Tel.: +82-2-820-5311; Fax: +82-2-825-1584

Received: 1 April 2019; Accepted: 1 May 2019; Published: 7 May 2019

Abstract: Piezoelectric and dielectric materials have attracted much attention for their functional device applications. Despite its excellent piezoelectric properties, the content of lead in piezoelectric materials should be restricted to prevent future environmental problems. Therefore, reduced lead content in piezoelectric materials with similar piezoelectric properties is favorable. In our research, piezoelectric materials with decreased lead content will be studied and discussed. Even though the lead content is decreased in $Bi(Mg_{0.5}Ti_{0.5})O_3$–$PbTiO_3$ ceramics, they show piezoelectric properties similar to that of lead zirconate titanate (PZT)-based materials. We believe this high piezoelectric behavior is related to the relaxation behavior of $Bi(Mg_{0.5}Ti_{0.5})O_3$–$PbTiO_3$ (BMT–PT) ceramics. In this study, $0.62Bi(Mg_{0.5}Ti_{0.5})O_3$–$0.38PbTiO_3$ ceramics were prepared by the conventional sintering process. These piezoelectric ceramics were sintered at varying temperatures of 975–1100 °C. Crystallinity and structural properties were analyzed and discussed. X-ray diffraction pattern analysis demonstrated that the optimal sintering temperature was around 1075 °C. A very high Curie temperature of 447 °C was recorded for 0.62BMT–0.38PT ceramics sintered at 1075 °C. For the first time, we found that the origin of the high Curie temperature, d_{33}, and the dielectric constant is the relaxation behavior of different dipoles in 0.62BMT–0.38PT ceramics.

Keywords: piezoelectric; BMT–PT; ceramic

1. Introduction

Since the discovery of Pb(Zr,Ti)O_3 (PZT) in the 1950s, it has been applied in piezoelectric transducers, actuators, and sensors because of its excellent piezoelectric properties [1]. PZT is also important in new high-precision switchable measurement methods, where dielectric properties are highly important. These methods compensate for environmental effects, voltage offset, frequency drift, and temperature influence, as previously reported in [2–4]. Due to its outstanding piezoelectric and mechanical properties, ceramic-based PZT actuators have received a great deal of attention in the industry. However, it is desirable to use materials with a low lead content to avoid environmental problems [5–8]. Therefore, there is a great need to discover piezoelectric materials that have a low lead content while still having favorable characteristics like those of PZT ceramics.

Recently, $BiFeO_3$-based (BF) perovskite materials have been intensively investigated, owing to their complex electric properties resulting from ferromagnetism and ferroelectricity [9–11]. For high temperature device applications, BF-based materials, $(1-x)BiMeO_3$–$xPbTiO_3$ (Me^{3+} = Fe, Sc, Mg, In, Y, Yb, Ga), are attractive. These materials contain metallic components in a distorted perovskite structure, which has a higher Curie temperature between the ferroelectric and paraelectric states than that of PZT ceramics [12]. In particular, bismuth-based perovskite system $Bi(Mg_{0.5}Ti_{0.5})O_3$–$PbTiO_3$ (BMT–PT) ceramics are expected to have a higher Curie temperature with relative polarization and piezoelectric charge coefficient. Since the origin of the $BiMeO_3$–$PTiO_3$ system is multiferroic material, it is expected to have weaker piezoelectric and dielectric properties than the PZT-based system. Despite the higher Curie temperature than that of the PZT system, lower piezoelectric and dielectric properties can be an obstacle for device applications. However, as a lead-containing material, BMT–PT piezoelectric material has similarly excellent piezoelectric properties as PZT, even though it has a low level of lead compared to PZT [13]. The multiferroic and relatively weak ferroelectric properties in the $BiMeO_3$–$PTiO_3$ system, which is usually observed in the electric field-dependent polarization process, can be the main obstacles for its future piezoelectric applications. However, as we have mentioned before, a relatively high piezoelectric charge coefficient of more than 200 pC/N with a higher Curie temperature range over 400 °C can improve its prospects for future multifunctional applications, including actuators and sensors.

In this study, we will prepare a BMT–PT system by optimizing the sintering temperature. The main advantages of a BMT–PT piezoelectric system are summarized in Table 1. As seen in Table 1, this system has high piezoelectric properties, a high Curie temperature, and a moderate price. However, as a representative lead content-reduced material, $BiScO_3$–$PbTiO_3$ also has a high piezoelectric coefficient and a high Curie temperature, but this material is very expensive and not feasible for application in electronic devices. The optimized composition of the BMT–PT system was selected through the phase diagram, and then the morphotropic phase boundary (MPB) was extracted to achieve the maximum piezoelectric properties [14]. In our assumption, we believe that $0.62Bi(Mg_{0.5}Ti_{0.5})O_3$–$0.38PbTiO_3$ ceramics have a mixture of rhombohedra and tetragonal structures. This assumption will be tested and discussed after X-ray diffraction (XRD) analysis. Therefore, the composition of $0.62Bi(Mg_{0.5}Ti_{0.5})O_3$–$0.38PbTiO_3$ was selected in this experiment [15]. Crystalline properties, including XRD patterns, and electrical properties, including piezoelectric and dielectric properties, will be investigated and discussed.

Table 1. Comparison of piezoelectric properties of $BiMeO_3$–$PbTiO_3$ and other lead content-reduced ceramics.

	Sintering Temperature	**Dielectric Permittivity**	**Curie Temperature**	**Piezoelectric Charge Coefficient**	**Cost**	**Reference**
$BiFeO_3$–$PbTiO_3$	1100 °C	500 for 0.3BF–0.7PT	500 °C	165 pC/N	Moderate	[12,16]
$BiScO_3$–$PbTiO_3$		1450 for 0.36BS–0.64PT	450 °C	460 pC/N	Very High	[12,17]
$BiInO_3$–$PbTiO_3$		600 for $0.1BiInO_3$–$0.9TiO_3$	541 °C		Moderate	[12,18]
$BiYbO_3$–$PbTiO_3$	1140 °C	650 for 0.1BY–0.9PT	590 °C	18 pC/N	Very High	[12,19]
BNT–PT $0.55Bi(Ni_{1/2}Ti_{1/2})O_3$–$0.45PbTiO_3$	1050 °C		400 °C	293 pC/N	Moderate	[12,20]
BMT–PT (our samples)	1075 °C	625 for 0.62BMT–0.38PT	447 °C	237 pC/N	Moderate	

2. Materials and Methods

The $0.62Bi(Mg_{0.5}Ti_{0.5})O_3$–$0.38PbTiO_3$ (0.62BMT–0.38PT) ceramics were prepared by a standard ceramic sintering technique using the blended oxides method. The raw materials, Bi_2O_3, $4MgCO_3 \cdot Mg(OH)_2 \cdot 5H_2O$, TiO_2, and PbO, were weighed and combined by ball-milling with a ZrO_2 ball in ethyl alcohol for 24 h. The blended powders were calcined at 900 °C for 2 h in a furnace. The desiccated powders were blended with polyvinyl alcohol (PVA) and compressed into a disk with a diameter of 10 mm and thickness of 1 mm. The sample was then sintered at various temperatures ranging from 975 to 1100 °C (975, 1000, 1025, 1050, 1075, and 1100 °C) for 2 h in a sealed alumina crucible to avoid loss of Bi_2O_3 and PbO due to sublimation. Silver electrodes were formed on both sides of the ceramic plates by screen printing. The poling process was performed in a silicon oil bath at

100 °C with an applied electric field of 1 kV/mm. The crystalline properties were analyzed by XRD using a Cu Kα radiation source (Bruker ARS), and the electric properties were analyzed by an HP 4294 impedance analyzer. Field emission scanning electron microscopy (FESEM) was used to examine the microstructure. To characterize its piezoelectric and dielectric properties, the piezoelectric charge coefficient and electric field-dependent polarization processes were performed, in this experiment, by employing the Sawyer–Tower method with 0.1 Hz.

3. Results and Discussion

Figure 1 shows the XRD patterns for the 0.62BMT–0.38PT systems, which were sintered at various temperatures. The specimens were sintered from 975 to 1100 °C in steps of 25 °C. Figure 2 displays the (001) XRD peaks of 0.62BMT–0.38PT piezoelectric ceramics according to the sintering temperature range. As shown in Figures 1 and 2, the XRD patterns show very weak variation in peak intensities and positions. This means that the degree of crystallization and d-space of lattice parameters were slightly changed after the sintering process. As observed in Figure 2, the (001) peak position of 0.62BMT–0.38PT ceramics were shifted to lower angles as the sintering temperature increased. It means that the lattice parameter c increased as the sintering temperature increased. Furthermore, it seems that 0.62BMT–0.38PT piezoelectric systems have good crystalline structure without a pyrochlore phase.

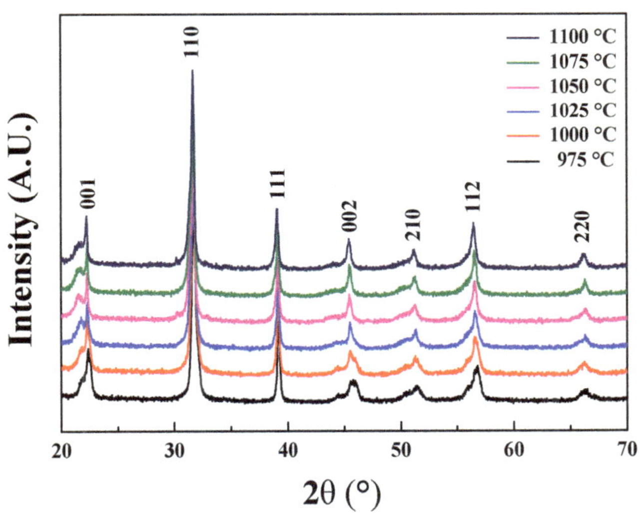

Figure 1. X-ray diffraction (XRD) patterns of $Bi(Mg_{0.5}Ti_{0.5})O_3$–$PbTiO_3$ ceramics according to sintering temperature.

Figure 3 depicts the degree of crystallization for the (001) direction in the 0.62BMT–0.38PT ceramic. The crystallization degree of (001) approached or even exceeded 10% at the sintering temperature of 1075 °C. This means that the distorted perovskite structure of 0.62BMT–0.38PT ceramics has a high degree of c-axis orientation as sintering temperature increases, which corresponds to the added thermal energy to form the crystalline structure. However, when the sintering temperature was increased beyond 1100 °C, the peak ratio of (001) decreased. We believe this maximum value of peak ratio (001) at 1075 °C and decreased peak ratio at 1100 °C are probably related with the crystallization degree. Therefore, we can argue that a sintering temperature of 1075 °C is the optimized sintering temperature owing to the increased (001) relative intensities.

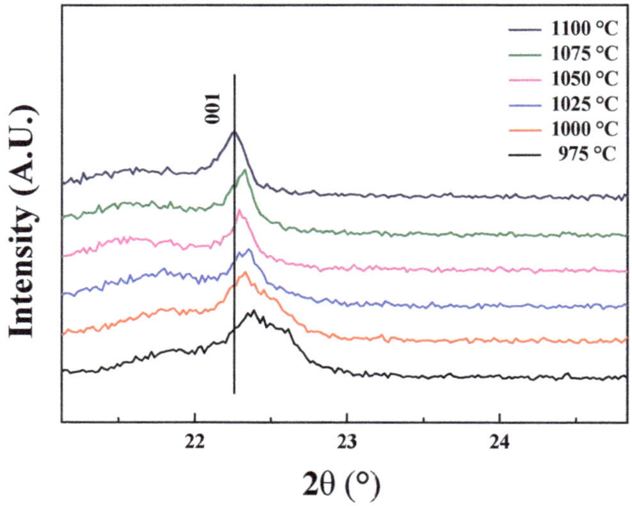

Figure 2. XRD (001) peak shifts of $Bi(Mg_{0.5}Ti_{0.5})O_3$–$PbTiO_3$ ceramics according to sintering temperature.

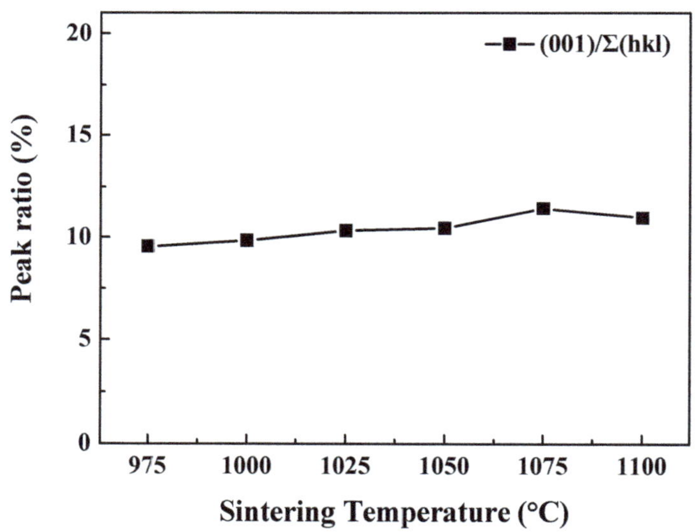

Figure 3. Peak ratio (001) of $Bi(Mg_{0.5}Ti_{0.5})O_3$–$PbTiO_3$ ceramics according to sintering temperature.

Figure 4 shows the FESEM images for the 0.62BMT–0.38PT ceramics, which were sintered at various temperatures. As shown in the FESEM images, the grain size of 0.62BMT–0.38PT ceramics increased as the sintering temperature increased. At a sintering temperature of 1075 °C, 0.62BMT–0.38PT ceramics showed a large grain size with a dense structure. Due to this highly dense structure, 0.62BMT–0.38PT ceramics that were sintered at 1075 °C showed the highest piezoelectric and dielectric properties among the specimens. However, as the sintering temperature of 1100 °C was attained, the grain shape was distorted, and the size was decreased.

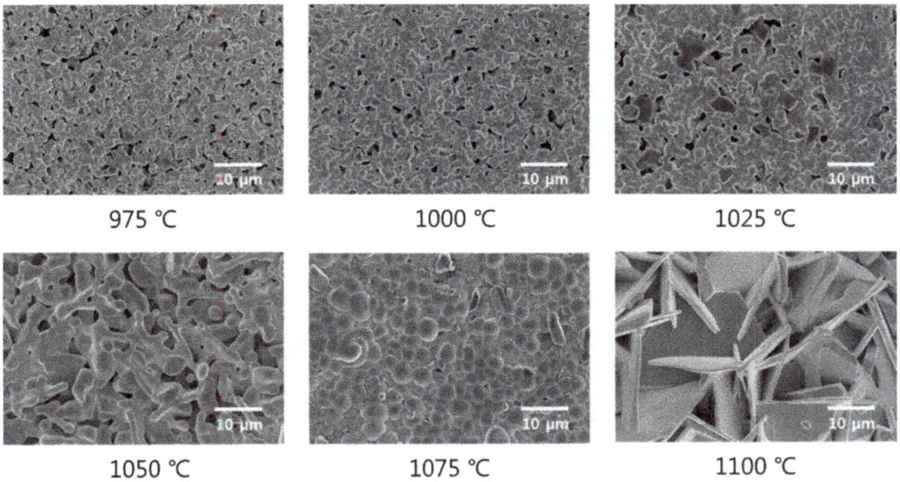

Figure 4. Field-emission scanning electron microscopy (FESEM) images of $Bi(Mg_{0.5}Ti_{0.5})O_3–PbTiO_3$ according to sintering temperature.

Figure 5 reveals the bulk and theoretical density of 0.62BMT–0.38PT ceramics depending on the sintering temperature. As the sintering temperature increased, the density increased up to 1075 °C and then decreased. This behavior is similar to that of grain size as observed in the FESEM images. As the sintering temperature increased, 0.62BMT–0.38PT ceramics became crystallized, therefore, the density was increased. However, as the sintering temperature was further increased up to 1100 °C, the density decreased. We believe that this decreased bulk density comes from the oversintered phase of 0.62BMT–0.38PT ceramics. It also seems that the bulk density of 0.62BMT–0.38PT ceramics correlated with the FESEM images, which is described in Figure 4.

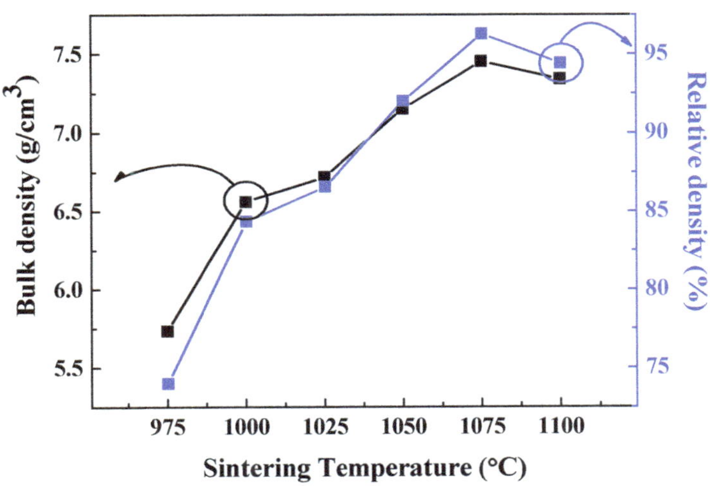

Figure 5. Bulk density and relative density of $Bi(Mg_{0.5}Ti_{0.5})O_3–PbTiO_3$ ceramics according to sintering temperature.

Figure 6 shows the frequency-dependent dielectric permittivity of 0.62BMT–0.38PT ceramics from 1 kHz to 1 MHz. Clearly, 0.62BMT–0.38PT ceramics sintered at the 1075 °C showed the highest dielectric permittivity of 625 at 1 kHz, whereas those sintered at 975 °C showed the lowest dielectric permittivity of 480 at 1 kHz.

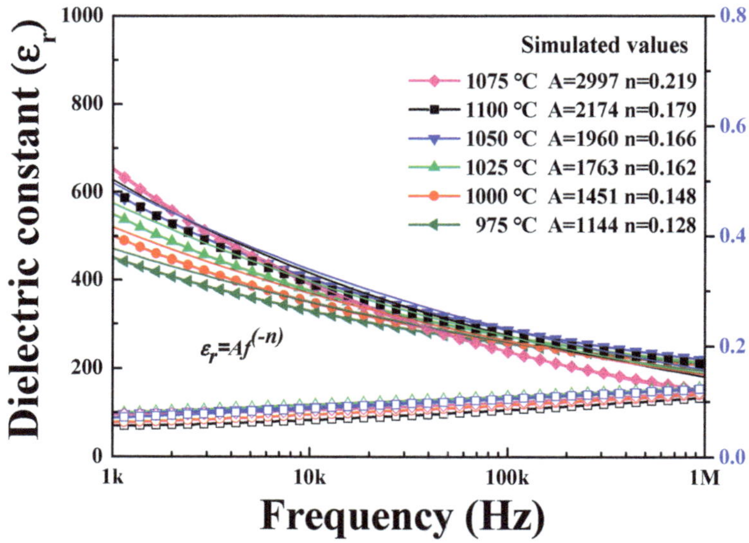

Figure 6. Frequency-dependent dielectric permittivity of $Bi(Mg_{0.5}Ti_{0.5})O_3$–$PbTiO_3$ ceramics according to sintering temperature.

It seems that 0.62BMT–0.38PT ceramics have high dielectric permittivity ranges at room temperature. The solid lines indicate the simulation fitting from the measured samples with the power law. The power law of $\varepsilon_r = Af^{-n}$ was employed to fit dielectric relaxation behavior [21]. Ceramics of 0.62BMT–0.38PT sintered at 1075 °C showed the highest exponent of 0.219 among the specimens. Compared to other specimens, this means that the variance of dielectric permittivity for 0.62BMT–0.38PT ceramics sintered at 1075 °C has the highest exponent of 0.219 and highest dielectric permittivity of 625. The high exponent value of 0.219 from the power law equation means that many different dipoles were involved in the relaxation process with continuous freezing out of dipoles with increasing frequency range. Therefore, 0.62BMT–0.38PT piezoelectric ceramics showed rapid change in the dielectric constant when increasing the frequency range. Owing to this high relaxation behavior, 0.62BMT–0.38PT ceramics sintered at 1075 °C have the highest piezoelectric properties. The dielectric permittivity of 0.62BMT–0.38PT ceramics sintered at 1075 °C also have the highest dielectric permittivity values.

Figure 7 shows the electric field-dependent polarizations for 0.62BMT–0.38PT ceramics with different sintering temperatures. As the sintering temperature was increased, the polarization increased. In particular, 0.62BMT–0.38PT ceramics sintered at 1075 and 1100 °C showed higher ferroelectric properties compared to other specimens. It seems that the higher the sintering temperature, the larger the polarization behavior. In the case of the electric field-dependent polarization case, sintering temperature dependencies are more clearly observed.

Figure 8 shows the piezoelectric charge coefficient and electromechanical coupling coefficient of 0.62BMT–0.38PT ceramics depending on the sintering temperature range. The piezoelectric charge coefficient and electromechanical coupling coefficient of 0.62BMT–0.38PT ceramics sintered at 1075 °C showed the highest values of 240 pC/N and 47%, respectively. These high piezoelectric charge coefficient and electromechanical coupling coefficient coincided well with the dense behavior observed in FESEM images and bulk densities.

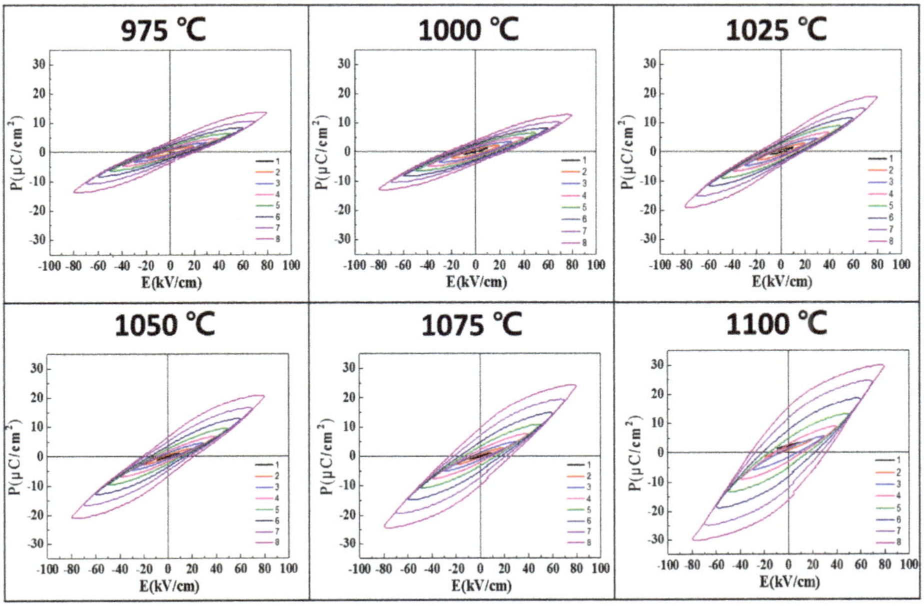

Figure 7. Electric field-dependent polarization properties of $Bi(Mg_{0.5}Ti_{0.5})O_3$–$PbTiO_3$ according to sintering temperature.

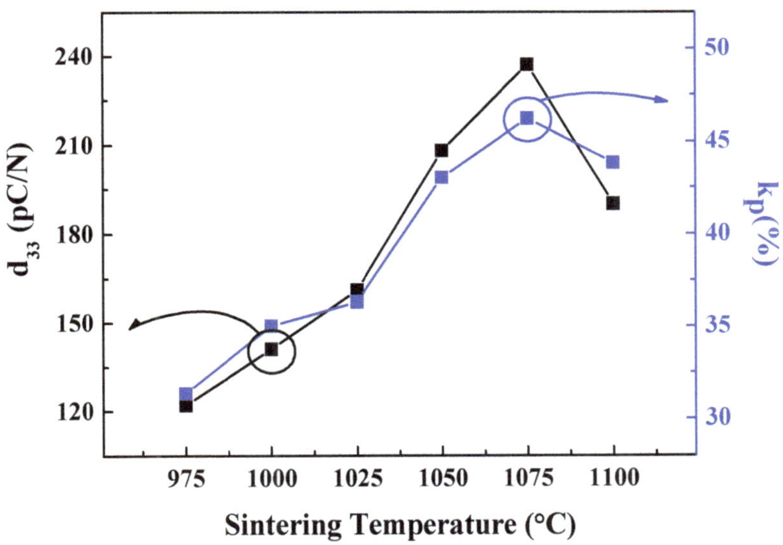

Figure 8. Piezoelectric coefficient d_{33} and k_p of $Bi(Mg_{0.5}Ti_{0.5})O_3$–$PbTiO_3$ according to sintering temperature.

Figure 9 displays the piezoelectric voltage coefficient and figure of merit (FoM) of 0.62BMT–0.38PT ceramics. As shown in Figure 9, the highest value of the piezoelectric voltage coefficient of 41×10^{-3} Vm/N and FoM of 9.8 pm^2/N were obtained for the 0.62BMT–0.38PT ceramics. Since the piezoelectric voltage coefficient was derived from the piezoelectric charge coefficient and dielectric constant, $g_{33} = d_{33}/\varepsilon_r$, we can expect that the piezoelectric voltage coefficient of 0.62BMT–0.38PT ceramics sintered at 1075 °C has the highest values of 41×10^{-3} Vm/N and FoM of 9.8 pm^2/N.

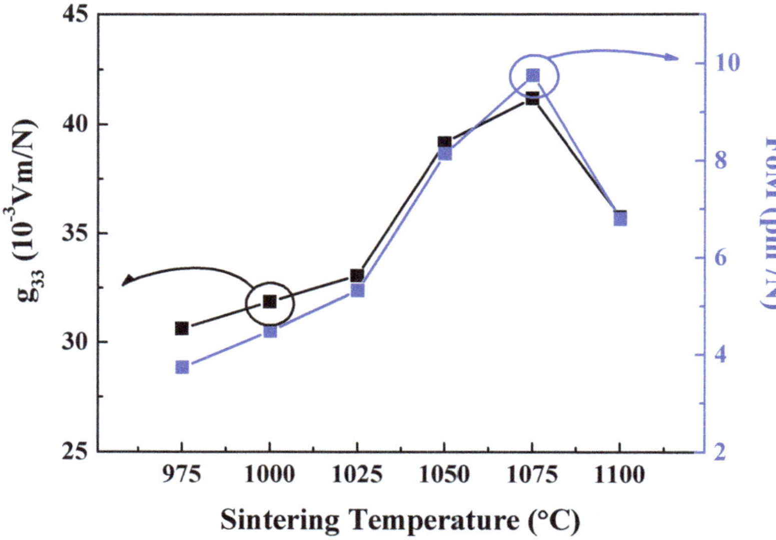

Figure 9. Piezoelectric voltage coefficient and figure of merit (FoM) of $Bi(Mg_{0.5}Ti_{0.5})O_3–PbTiO_3$ according to sintering temperature.

Figure 10 shows the temperature-dependent dielectric permittivity ε_r of the 0.62BMT–0.38PT ceramic sintered from 975 to 1100 °C. At sintering temperatures of 975, 1000, 1025, 1050, 1075, and 1100 °C, the Curie temperature was 346.9, 374.5, 386.5, 389.2, 447.2, and 428.5 °C, respectively. As the sintering temperature increased from 975 to 1075 °C, the Curie temperature increased from 346.9 to 447.2 °C. As the sintering temperature reached 1100 °C, the Curie temperature suddenly decreased. Compared to other specimens, BMT–PT ceramics sintered at 975 °C had the lowest Curie temperature of 346.9 °C, whereas those sintered at 1075 °C had the highest Curie temperature of 447.2 °C.

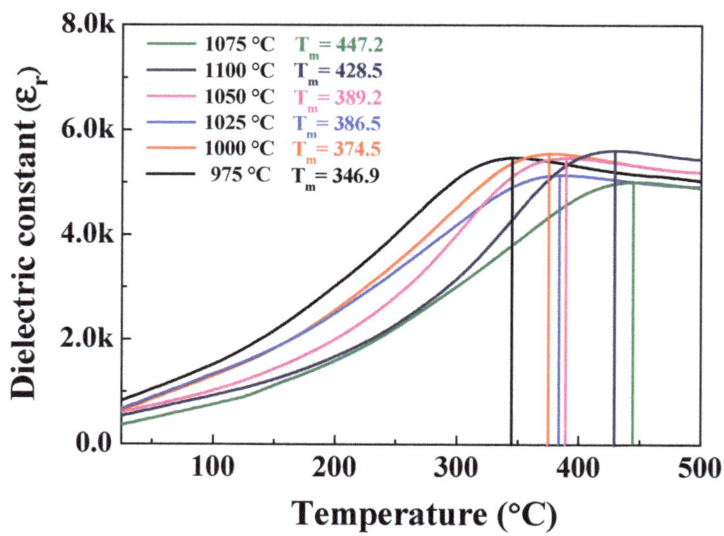

Figure 10. Temperature-dependent relative dielectric constant ε_r of $Bi(Mg_{0.5}Ti_{0.5})O_3–PbTiO_3$ ceramic according to sintering temperature.

4. Conclusions

In this research, the piezoelectric and dielectric properties of 0.62BMT–0.38PT ceramics with different sintering temperatures were investigated and discussed. A very high Curie temperature of 447 °C was recorded for 0.62BMT–0.38PT ceramics sintered at 1075 °C. 0.62BMT–0.38PT piezoelectric ceramics have a reduced lead content and high piezoelectric charge coefficient of 237 pC/N and can, therefore, be used in piezoelectric applications where the environment is harsh. Even though 0.62BMT–0.38PT ceramics had a lower lead content compared to $Pb(Zr_{0.5}Ti_{0.5})O_3$-based conventional piezoelectric ceramics, 0.62BMT–0.38PT piezoelectric ceramics still have excellent piezoelectric and dielectric properties with a high Curie temperature. We believe that the high piezoelectric properties of 0.62BMT–0.38PT are related to relaxation behavior. We believe 0.62BMT–0.38PT ceramics, with their lower lead content, can act as alternative materials for lead-based piezoelectric materials. Reducing the lead content of piezoelectric materials is expected to help not only by preventing environmental pollution on Earth but, also, by being more useful as a piezoelectric material because of its excellent characteristics compared to lead-free piezoelectric material. For example, 0.62BMT–0.38PT ceramics can be applied in piezoelectric transducers or as a sintering aid, using ferroelectricity, that does not change at a high temperature.

Author Contributions: M.Y.P. performed the experiments and wrote the paper; J.-H.J. made some additional experiment and analysis and J.-H.K. analyzed experimental data and wrote the paper.

Funding: This research was supported by the MSIP, Korea, under the ITRC support program (IITP-2018-2014-1-00636) supervised by the IITP and supported by the Human Resources Development (No. 20184030202070) of the Korea Institute of Energy Technology Evaluation and Planning (KETEP) grant funded by the Korea government Ministry of Trade, Industry and Energy.

Conflicts of Interest: The authors declare no conflict of interest.

References

1. Wang, L.; Song, T.K.; Lee, S.C.; Cho, J.H.; Sakka, Y. Effect of $Bi(B)O_3$ perovskite substitution on enhanced tetragonality and ferroelectric transition temperature in $Pb(Zr,Ti)O_3$ ceramics. *Mater. Chem. Phys.* **2011**, *129*, 322–325. [CrossRef]
2. Matko, V.; Milanovic, M. Temperature-compensated capacitance-frequency converter with high resolution. *Sens. Actuators A* **2014**, *220*, 262–269. [CrossRef]
3. Matko, V. Next Generation AT-Cut Quartz Crystal Sensing Devices. *Sensors* **2011**, *11*, 4474–4482. [CrossRef] [PubMed]
4. Nie, J.; Liu, J.; Li, N.; Meng, X. Dew point measurement using dual quartz crystal resonator sensor. *Sens. Actuators B* **2017**, *246*, 792–799. [CrossRef]
5. Li, J.-F.; Wang, K. Ferroelectric and Piezoelectric Properties of Fine-Grained $Na_{0.5}K_{0.5}NbO_3$ Lead-Free Piezoelectric Ceramics Prepared by Spark Plasma Sintering. *J. Am. Ceram. Soc.* **2006**, *89*, 706–709. [CrossRef]
6. Takenaka, T.; Nagata, H. Current status and prospects of lead-free piezoelectric ceramics. *J. Eur. Ceram. Soc.* **2005**, *25*, 2693–2700. [CrossRef]
7. Baettig, P.; Schelle, C.F.; LeSar, R.; Waghmare, U.V.; Spaldin, N.A. Theoretical Prediction of New High-Performance Lead-Free Piezoelectric. *Chem. Mater.* **2005**, *17*, 1376–1380. [CrossRef]
8. Saleem, M.; Hwan, L.D.; Kim, I.-S.; Kim, M.-S.; Maqbool, A.; Nisar, U.; Pervez, S.A.; Farooq, U.; Farooq, M.U.; Khalil, H.M.W.; et al. Revealing of Core Shell Effect on Frequency-Dependent Properties of Bi-based Relaxor/Ferroelectric Ceramic Composites. *Sci. Rep.* **2018**, *8*, 14146. [CrossRef] [PubMed]
9. Yabuta, H.; Shimada, M.; Watanabe, T.; Hayashi, J.; Kubota, M.; Miura, K.; Fukui, T.; Fujii, I.; Wada, S. Microstructure of $BaTiO_3$-$Bi(Mg_{1/2}Ti_{1/2})O_3$-$BiFeO_3$ Piezoelectric Ceramics. *Jpn. J. Appl. Phys.* **2012**, *51*, 09LD04. [CrossRef]
10. Leontsev, S.O.; Eitel, R.E. Dielectric and Piezoelectric Properties in Mn-Modified (1-x)$BiFeO_3$-x$BaTiO_3$ Ceramics. *J. Am. Ceram. Soc.* **2009**, *92*, 2957–2961. [CrossRef]
11. Jiang, M.; Liu, X.; Chen, G.; Zhou, C. Dielectric and piezoelectric properties of $LiSbO_3$ doped 0.995 $K_{0.5}Na_{0.5}NbO_3$-0.005$BiFeO_3$ piezoelectric ceramics. *Mater. Lett.* **2009**, *63*, 1262–1265. [CrossRef]

12. Rai, R.; Kholkin, A.L.; Sharma, S. Multiferroic properties of $BiFeO_3$ doped $Bi(MgTi)O_3$-$PbTiO_3$ ceramic system. *J. Alloys Comps.* **2010**, *506*, 815–819. [CrossRef]
13. Kwon, Y.-H.; Ahn, J.-H.; Koh, J.-H. The Dielectric Properties of $Bi(Mg,Ti)O_3$-$PbTiO_3$ Ceramics. *J. Nanosci. Nanotechnol.* **2016**, *16*, 12927–12929. [CrossRef]
14. Randall, C.A.; Eitel, R.; Jones, B.; Shrout, T.R. Investigation of a high Tc piezoelectric system: $(1-x)Bi(Mg_{1/2}Ti_{1/2})O_3$-$(x)PbTiO_3$. *J. Appl. Phys.* **2014**, *95*, 3633–3639. [CrossRef]
15. Zhang, Q.; Li, Z.; Li, F.; Xu, Z.; Yao, X. Temperature Dependence of Dielectric/Piezoelectric Properties of $(1-x)Bi(Mg_{1/2}Ti_{1/2})O_3$-$xPbTiO_3$ Ceramics with an MPB Composition. *J. Am. Ceram. Soc.* **2010**, *93*, 3330–3334. [CrossRef]
16. Zhao, H.; Hou, Y.; Yu, X.; Zheng, M.; Zhu, M. Construction of high Tc $BiScO_3$-$BiFeO_3$-$PbTiO_3$ and its enhanced piezoelectric properties by sintering in oxygen atmosphere. *J. Appl. Phys.* **2018**, *124*, 194103. [CrossRef]
17. Zhang, S.; Alberta, E.F.; Eitel, R.E.; Randall, C.A.; Shrout, T.R. Elastic, Piezoelectric, and Dielectric Characterization of Modified $BiScO_3$-$PbTiO_3$ Ceramics. *IEEE Trans. Ultrason. Ferroelectr. Freq. Control* **2005**, *52*, 2131–2139. [CrossRef] [PubMed]
18. Qin, B.-Q.; Chen, Y.; Jiang, Y.; Jiang, Y.; Yue, X.; Xiao, D.; Zhu, J. Preparation and Characterization of $(1-x)BiInO_3$-$xPbTiO_3$ ceramics. In Proceedings of the 2007 Sixteenth IEEE International Symposium on the Applications of Ferroelectrics, Nara, Japan, 27–31 May 2007; pp. 616–617.
19. Gao, F.; Hong, R.; Liu, J.; Li, Z.; Cheng, L.; Tian, C. Phase formation and characterization of high Curie temperature $xBiYbO_3$–$(1-x)PbTiO_3$ piezoelectric ceramics. *J. Eur. Ceram. Soc.* **2009**, *29*, 1687–1693.
20. Jiang, S.; Zhu, Z.; Zhang, L.; Xiong, X.; Yi, J.; Zeng, Y.; Liu, W.; Wang, Q.; Han, K.; Zhang, G. Electrical properties of $Bi(Ni_{1/2}Ti_{1/2})O_3$–$PbTiO_3$ high-T_C piezoelectric ceramics fabricated by the microwave sintering process. *Mater. Sci. Eng. B Solid State Mater. Adv. Technol.* **2014**, *179*, 36–40. [CrossRef]
21. Weron, K.; Stanislavsky, A.; Jurlewicz, A.; Meerschaert, M.M.; Scheffler, H.-P. Clustered continuous-time random walks: diffusion and relaxation consequences. *Proc. R. Soc. A* **2012**, *468*, 1615–1628. [CrossRef] [PubMed]

sensors

Article

Thermal Wave Scattering by an Elliptic Subsurface Hole Buried in a Block, Based on the Non-Fourier Equation

Chuanping Zhou [1,2] and Ban Wang [1,2,*]

1 College of Electrical Engineering, Zhejiang University; Hangzhou 310027, China; zhouchuanping@126.com
2 School of Mechanical Engineering, Hangzhou Dianzi University; Hangzhou 310018, China
* Correspondence: bigban@zju.edu.cn

Received: 20 March 2019; Accepted: 15 April 2019; Published: 19 April 2019

Abstract: With the application to engineering practice, the study of the scattering of thermal waves using innovative and comprehensive methods is becoming increasingly important. The thermal wave scattering by an elliptic subsurface hole in a block with two boundaries is discussed based on the non-Fourier heat conduction equation, employing the complex function method and the conformal mapping method, and a general solution for the thermal wave scattering is given. The numerical results of temperature distributions around a subsurface hole are presented and the effects of geometrical and physical parameters on the temperature distributions were analyzed. The wave number, the shape and position of the hole, the scale of the block, and the frequency of the heat load were found to have great effects on distributions and variations of temperature. The findings of this study could be helpful in providing better understandings of infrared thermal wave imaging, the physical inverse problem, and the evaluation of internal holes in materials.

Keywords: non-Fourier law; two boundaries; elliptic hole; thermal wave scattering; complex function method

---

1. Introduction

Subsurface defects can be detected and evaluated by tuning the temperature in the thermal wave field, since the frequency and amplitude of ultra-short laser pulses can be controlled [1–3]. It is an accessible way to get the subsurface micro-structure information and to realize thermal wave detection according to the real-time measurement of the temperature field on the solid surface. When the temperature increases quickly, the heat conduction process in solids should be described by a hyperbolic equation, which means that only wave equations can be used to illustrate the features and properties of the heat conduction process [4–6]. It has been proven that results calculated based on the non-Fourier equation are more consistent with engineering practices [7–9]. In most cases, the classical Fourier heat conduction law is an excellent description of heat conduction physics. In practical engineering, heat sources such as lasers and microwaves with extremely short durations, very high frequencies, or quite high heat-flux densities are widely used. The non-Fourier heat conduction phenomenon has been found in these media. Many researchers have attached much importance to the potential practical values of non-Fourier heat conduction in many applications, and non-Fourier heat conduction has become one of the hotspots in the field of heat transfer.

It can also evaluate the applicable conditions for the classic heat conduction equation, which has great prospects in theory and engineering for analyzing the thermal wave multiple scattering and temperature distribution with the hyperbolic heat conduction equation.

Most recently, with the advantages of non-contact accuracy and sensitivity, infrared thermography is frequently used in the nondestructive evaluation of solid materials containing subsurface inclusions [3,5,10]. Thermal wave detection technology using the infrared lock-in thermography method adopts the heat source of the single–frequency modulation intensity of the sine law [11]. The scattering of thermal waves can also have applications in other fields, such as thermal recovery from heavy oil reservoirs [12,13] and the sensing property of defective temperature sensors. Consequently, investigations into the scattering of thermal waves are becoming increasingly important. Nevertheless, most studies on the scattering of thermal waves are aimed at infinite body models or semi-infinite body models with a single structure boundary [14–16]. These models are usually applied to the structures in which the scale along the thermal perturbation propagation direction is much larger than the thermodynamics thickness [17]. Besides, the shapes of scattered bodies are almost always assumed to be circular and the effects resulting from the finiteness of other scales in project practice are also ignored [16,18]. This means that the applications of the theories outlined above have limitations in engineering practice. Therefore, it is necessary to pursue innovative and comprehensive methods.

The main objective of this paper was to investigate the multiple scattering of thermal waves by a subsurface, non-circular hole in an infinite block with two boundaries, based on the non-Fourier heat conduction hyperbolic equation. The problem can be reduced to the solution of an infinite system of algebraic equations. Compared to previous theoretical study, the outstanding novelties of this study are given as follows. The non-Fourier heat conduction equation can be used to calculate the temperature and thermal stress of materials and structures in extreme environments, but the traditional Fourier equation is not invalid. The Fourier equation is applicable in an infinite body, but the calculation error is larger in structures with boundaries. Our theory can be used to calculate the thermal wave scattering around non-circular subsurface holes. First, the wave equation of heat conduction and its general solution are described. Then, as examples, numerical results of thermal wave scattering in the block with an elliptical hole are computed. Lastly, the effects of incident wave number, physical dimensions, and parameters on the temperature distribution are analyzed, especially those of the incident wave number. Given proper mapping functions, this paper brings about an important significance in solving the problem of the scattering of thermal waves by holes in any shapes [19].

2. Wave Equation of Heat Conduction and Its General Solution

An infinite block with a hole was considered, and is depicted in Figure 1. An elliptic hole was embedded in an adiabatic solid. An ultra-short laser pulse, modulated at a frequency f, hit the surface of the heated materials along the z direction. Providing the temperature field distribution along the z direction is homogeneous, the temperature is determined by x and y only.

Based on the non-Fourier law of heat conduction, the governing equation of temperature in the solid can be written as:

$$\nabla^2 T = \frac{1}{c^2}\frac{\partial^2 T}{\partial t^2} + \frac{1}{D}\frac{\partial T}{\partial t}, \tag{1}$$

where ∇^2 is the Laplace operator, and $\nabla^2 = \frac{\partial^2}{\partial x^2} + \frac{\partial^2}{\partial y^2}$; λ, c_p, and ρ are the thermal conductivity, the specific heat at constant pressure, and the density, respectively. The thermal diffusivity is $D = \lambda/\rho c_p$, $c = \sqrt{D/\tau}$ is the thermal wave propagation velocity, τ is the heat relaxation time, and T is the temperature of the solid medium.

This divides the solution of a wave propagation problem into both a time domain solution and a frequency domain solution. By means of the Fourier transform, the two kinds of solutions can be transformed. To detect the position, direction, and size of flaws in the materials, employing the phase and amplitude of thermal response, as well as wave number and the space attenuation coefficient, periodic thermal waves are excited in the materials.

According to the Fourier decomposition theorem, the periodic heat conduction process is considered to be the composition of several simple harmonic waves. Periodic steady conduction can be analyzed in light of the format standardization of wave theory.

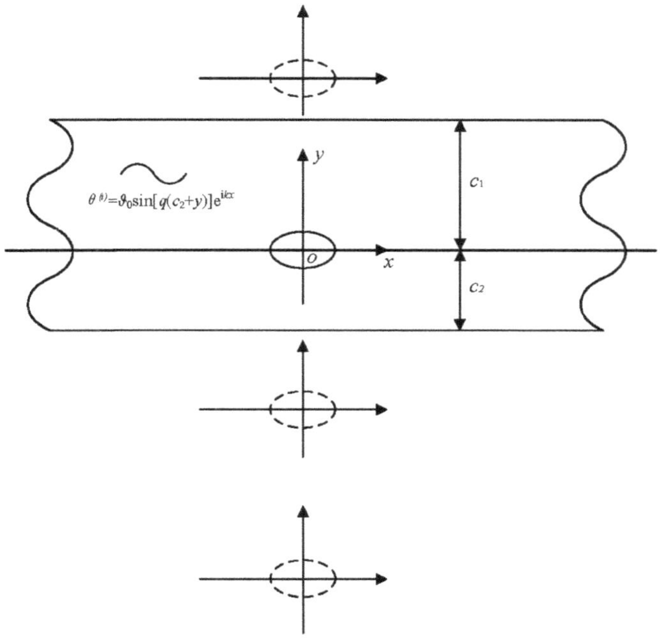

Figure 1. Laser pulse heating on specimen.

As the solution to Equation (1) is investigated, the temperature field can be described as

$$T = T_m + \mathrm{Re}[\vartheta \exp(-\mathrm{i}\omega\, t)] \tag{2}$$

where ϑ is the temperature amplitude, which should meet with the Helmholtz equation as follows:

$$\nabla^2\vartheta + \kappa^2\vartheta = 0 \tag{3}$$

In which Re denote real part; T_m is the ambient mean temperature; ω is the incident frequency, $\omega = 2\pi f$; $\mathrm{i} = \sqrt{-1}$ is the imaginary unit; κ is the wave number of complex variables, $\kappa = \left(\frac{\omega^2}{c^2} + \mathrm{i}\frac{\omega}{D}\right)^{1/2} = \alpha + \mathrm{i}\beta$; and $\alpha,\ \beta$ are the wave number and the absorption coefficient of thermal waves, respectively. Without loss of generality, after normalizing and taking $\alpha < 0,\ \beta < 0$, we can obtain

$$\alpha = \sqrt{\frac{1}{2}[\sqrt{\frac{\omega^4}{c^4} + \frac{\omega^2}{D^2}} + \frac{\omega^2}{c^2}]} = \sqrt{\sqrt{\frac{1}{4}k^4 + \frac{1}{\mu^4}} + \frac{1}{2}k^2}$$

$$\beta = \sqrt{\frac{1}{2}[\sqrt{\frac{\omega^4}{c^4} + \frac{\omega^2}{D^2}} - \frac{\omega^2}{c^2}]} = \sqrt{\sqrt{\frac{1}{4}k^4 + \frac{1}{\mu^4}} - \frac{1}{2}k^2}$$

Here, k is the wave number of thermal waves without diffusive effect. When the propagating speed of the thermal wave is $c \to \infty$, the non-Fourier equation degenerates into the classical Fourier equation, and then

$$\alpha \to \sqrt{\omega/2D} = \sqrt{\pi f/D} = 1/\mu,\ \beta \to \sqrt{\omega/2D} = 1/\mu$$

while $\kappa = \alpha + \mathrm{i}\beta \to (1+\mathrm{i})\sqrt{\omega/2D} = (1+\mathrm{i})/\mu$.

Consequently, there exists the wave motion with the form $\vartheta e^{-\mathrm{i}\omega t} = Ae^{-\beta x}e^{\mathrm{i}(\alpha x - \omega t)}$, which denotes the propagating thermal waves with their vibration amplitude attenuating in the z direction.

Here, a complex function method is employed, with the complex variable

$$z = x + \mathrm{i}y,\ \bar{z} = x - \mathrm{i}y; \tag{4}$$

that is,

$$x = \frac{1}{2}(z + \bar{z}),\ y = \frac{1}{2\mathrm{i}}(z - \bar{z}). \tag{5}$$

In this way, Equation (3) is converted into the form:

$$\frac{\partial^2 \vartheta}{\partial z \partial \bar{z}} + \left(\frac{\kappa}{2}\right)^2 \vartheta = 0. \tag{6}$$

The general solution to the scattered field of thermal waves in the solid medium determined by Equation (6) can be described as

$$\vartheta = \sum_{n=-\infty}^{\infty} A_n H_n^{(1)}(\kappa r) \mathrm{e}^{\mathrm{i}n\theta} = \sum_{n=-\infty}^{\infty} A_n H_n^{(1)}(\kappa|z|) \left\{\frac{z}{|z|}\right\}^n, \tag{7}$$

where A_n are the mode coefficients resulting from the subsurface cylinder hole and are determined by the boundary conditions, and $H_n^{(1)}(\cdot)$ is the Hankel function of the first kind.

3. The General Solution for the Scattering of Thermal Waves by Subsurface Holes Buried in an Infinite Block

In the process of solving the problem of the scattering of thermal waves by elliptically shaped holes, the conformal mapping method is a feasible way to provide access to meet with the boundary conditions of the hole. The boundaries of non-circular holes in the z plane can be mapped into a unit circle in the ζ plane by using the following equation:

$$z = \Omega(\zeta). \tag{8}$$

As the polar coordinates system is mentioned, there is $z = re^{\mathrm{i}\varphi}$ in the z plane, while $\zeta = \rho e^{\mathrm{i}\theta}$ is in the ζ plane. In this way, the general solution to the scattered field of thermal waves by non-circular holes in the ζ plane can be described as:

$$\vartheta = \sum_{n=-\infty}^{\infty} A_n H_n^{(1)}\big(\kappa|\Omega(\zeta)|\big) \left\{\frac{\Omega(\zeta)}{|\Omega(\zeta)|}\right\}^n. \tag{9}$$

For an ellipse with a major radius r_1 and a minor radius r_2, the conformal mapping function is denoted as:

$$z = \Omega(\zeta) = \frac{a}{1+\varepsilon}\left(\zeta + \frac{\varepsilon}{\zeta}\right), \tag{10}$$

where $a = (r_1 + r_2)/2, \varepsilon = (r_1 - r_2)/(r_1 + r_2)$.

4. The Incidence of Thermal Waves and Total Wave Field

Thermal waves can be generated at the fore end of the block by a laser beam with a modulated ultra-short pulse. It propagates along the positive x direction. Based on the interference theory of wave field, the temperature distribution in the infinite block is described as:

$$\vartheta = f(y) \exp[\mathrm{i}(px - \omega t)], \tag{11}$$

which needs to meet with Equation (2); then, Equation (12) is accessible as:

$$f(y) = A\cos(q\,y) + B\sin(qy), \tag{12}$$

where p and q are longitudinal wave number and lateral wave number, respectively, and $p^2 = \kappa^2 - q^2$. Meanwhile, the upper and lower signs refer to

$$f(c_1)\exp(ipx) = 0;\ f(-c_2)\exp(ipx) = 0. \tag{13}$$

In this way, the lateral wave number is derived:

$$q = \frac{n\pi}{c_1 + c_2}\ (n = 0, 1, 2 \cdots \infty) \tag{14}$$

Substituting Equation (14) into Equation (6), the following equation is obtained:

$$\vartheta = B\frac{\sin[q(c_2 + y)]}{\cos(qc_2)}\exp[\mathrm{i}(px - \omega t)]. \tag{15}$$

The thermal wave can be described as:

$$\vartheta^{(i)} = \vartheta_0 \sin[q(c_2 + y)]\mathrm{e}^{\mathrm{i}px-\omega t} = \vartheta_0 \sin[q(c_2 + y)] \sum_{n=-\infty}^{\infty} \mathrm{i}^n J_n(pr)\mathrm{e}^{\mathrm{i}n\theta}\mathrm{e}^{-\mathrm{i}\omega t}, \tag{16}$$

where ϑ_0 is the temperature amplitude of the incidence thermal wave, that is, excess temperature $\vartheta_0 = T_0 - T_m$; $q = \pi/(c_1 + c_2)$, p is the wave number along the x direction, and $J_n(\cdot)$ is the Bessel function.
Correspondingly, under the adiabatic conditions, the thermal wave can be described as:

$$\vartheta^{(i)} = \vartheta_0 \cos[q(c_2 + y)]\mathrm{e}^{\mathrm{i}px}\mathrm{e}^{-\mathrm{i}\omega t}. \tag{17}$$

Taking the multiple scattering by the solid boundary when $y = c_1$ and $y = -c_2$ $(c_1 < 0, c_2 < 0)$ into consideration, the scattered field of thermal waves by non-circular holes can be described in the polar coordinates system as:

$$\vartheta^{(s)} = \sum_{n=-\infty}^{\infty} A_n H_n^{(1)}(\kappa r)\mathrm{e}^{\mathrm{i}n\varphi} + \sum_{m=1}^{\infty}\sum_{n=-\infty}^{\infty} A_n H_n^{(1)}(\kappa r_m)\mathrm{e}^{\mathrm{i}n\varphi_m}. \tag{18}$$

By means of the complex function method, we can obtain the complex function expression of Equation (18) in Appendix A.

Consequently, the total wave field is composed of the incidence and scattering fields, which is conveyed by Equation (20):

$$\vartheta^{(t)} = \vartheta^{(i)} + \vartheta^{(s)}. \tag{19}$$

5. Mode Coefficient of Waves and Temperature Distribution

Substituting Equation (19) into the Dirichlet boundary condition of heat conduction, the following equation is obtained:

$$\sum_{n=-\infty}^{+\infty} {}_nX_n =, \tag{20}$$

the expression of which is shown in Appendix B.

When Equation (20) is multiplied by $\mathrm{e}^{-\mathrm{i}s\theta}$ at both ends, and by evaluating its integrals from $-\pi$ to π, the infinite algebraic equation is described as:

$$\sum_{n=-\infty}^{+\infty} {}_{ns}X_n =_s\ (n = s = 0,\ \pm1,\ \pm2 \cdots). \tag{21}$$

Equation (21) is the infinite algebraic equation determining the mode coefficients of thermal waves.

Let the major radius r_1 of the subsurface ellipse hole be the characteristic length, and the amplitude of the incident temperature be $|\vartheta_0|$. During computation, the following dimensionless variables were adopted: the ratios between the major radius and the minor radius of ellipse hole were $r_1/r_2 = 3/4, 1.0, 4/3$, respectively. The wave number of the non-diffusive propagating waves was $ka = 0.01 - 2.0$, the relative length of thermal diffusion was $\mu/a = 0.1 - 5.0$, the distance from the center of the ellipse to the upper and lower signs of the block was $c_2/a = 5.0 - 10.0$, $c_1/a = 5.0 - 30.0$, and the ratio of the excess temperature was ϑ/ϑ_0.

Thus, the expression of the temperature distribution at the surface is written as:

$$\begin{aligned}\vartheta = \vartheta_0 \sin[q(c_2 + y)]\mathrm{e}^{\mathrm{i}p(x+b)} + \sum_{n=-\infty}^{\infty} A_n H_n^{(1)}(\kappa|z|)\left\{\frac{z}{|z|}\right\}^n \\ + \sum_{n=-\infty}^{\infty} A_n \sum_{l=1}^{4} \sum_{m=1}^{\infty} (-1)^l [H_n^{(1)}(\kappa|z - z_{lm}|)\left\{\frac{z-z_{lm}}{|z-z_{lm}|}\right\}^{(-1)^l n}\end{aligned} \tag{22}$$

6. Numerical Examples

Figures 2 and 3 illustrate the temperature distributions of thermal scattering by a subsurface hole in an infinite block. It can be seen that the peak temperature on the surface of the material appeared right ahead of the elliptical hole. Note that when the wave number of the non-diffusive incident wave was small (e.g., $ka < 0.3$), the wave characteristics of heat conduction were weak. In this case, the temperature distribution was approximately the same as in the results from the classical heat conduction equation, which implies the applicability of the classical heat conduction equation. Furthermore, it was observed that the temperature varied slowly when the wave number was small (e.g., $ka = 0.1$). Otherwise, the temperature varied drastically.

Figures 4 and 5 illustrate the temperature distributions of thermal scattering corresponding to different eccentricities of the ellipse. Specifically, $\varepsilon = 0$, $|\varepsilon| = 1/7$, and $|\varepsilon| = 1/3$ represent the cases of a circular hole, a small eccentricity ratio, and a large eccentricity ratio, respectively. It is shown that the temperature corresponding to the case of a circular hole was the highest among the three at any point. In addition, when the major radius of the ellipse was along the lateral scale (i.e., $\varepsilon < 0$), the temperature fluctuated mildly and its peak value dropped slowly, and vice versa ($\varepsilon < 0$). Furthermore, only when the center of the hole was on the median of the block was the temperature distribution symmetrical.

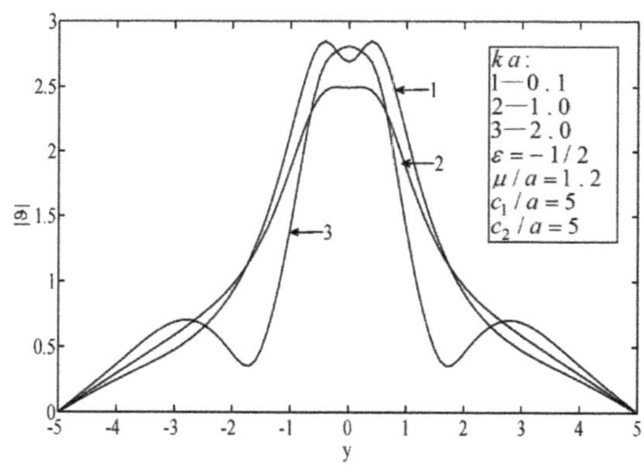

Figure 2. Temperature on the surfaces of the block.

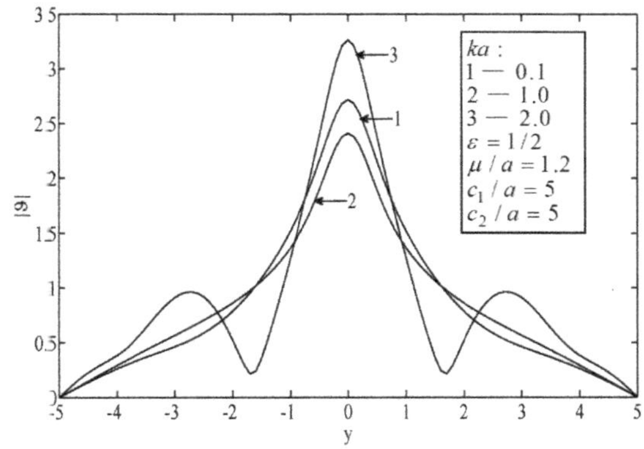

Figure 3. Temperature on the surfaces of the block.

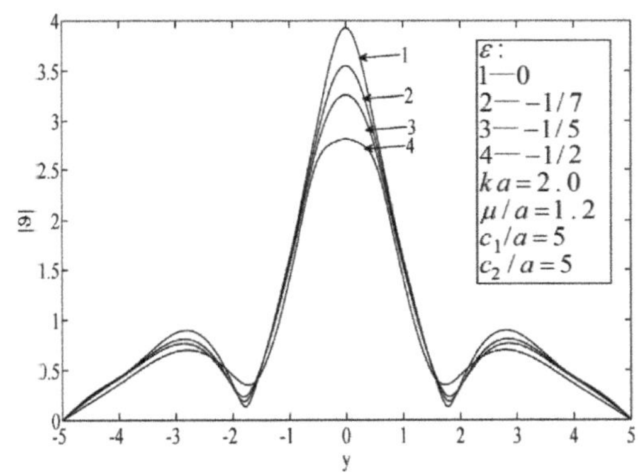

Figure 4. Temperature on the surfaces of the block.

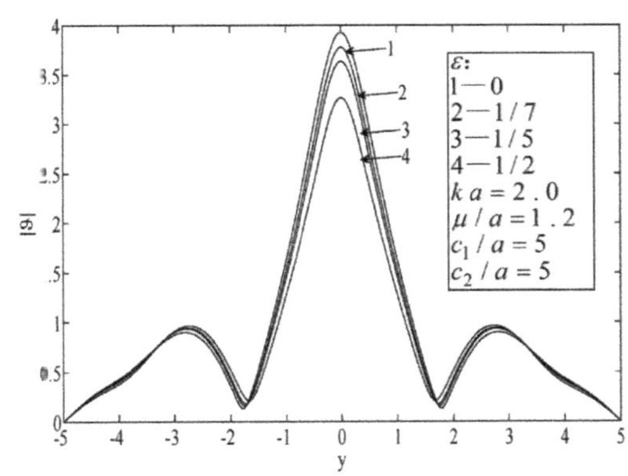

Figure 5. Temperature on the surfaces of the block.

As shown in Figures 6–8, the peak temperature appeared at the median of the block when the hole was eccentric to the block, which indicates that the temperature directly in front of the hole decreased. When the size of the block was small, the peak temperature on the surface was relatively high.

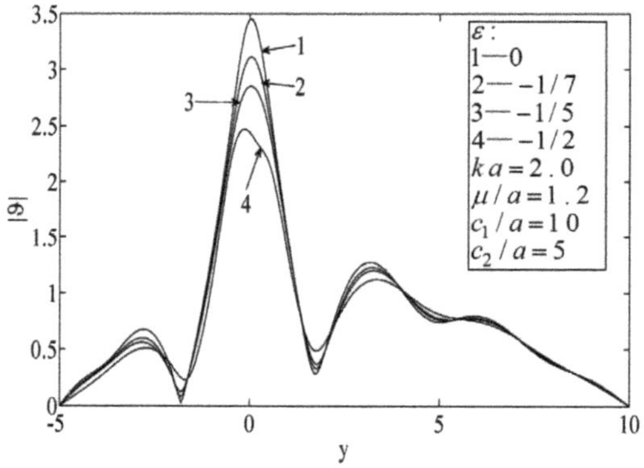

Figure 6. Temperature on the surfaces of the block.

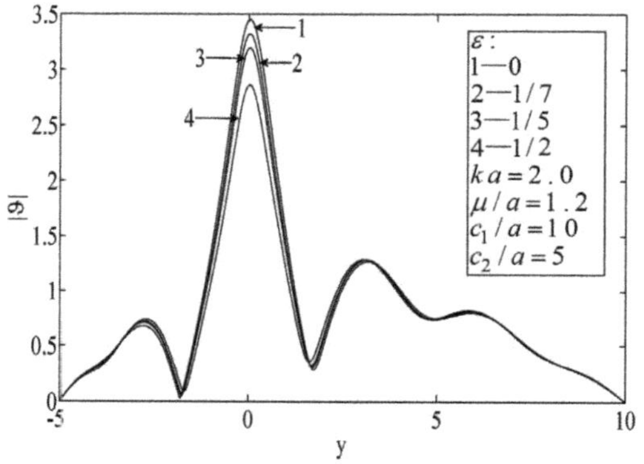

Figure 7. Temperature on the surfaces of the block.

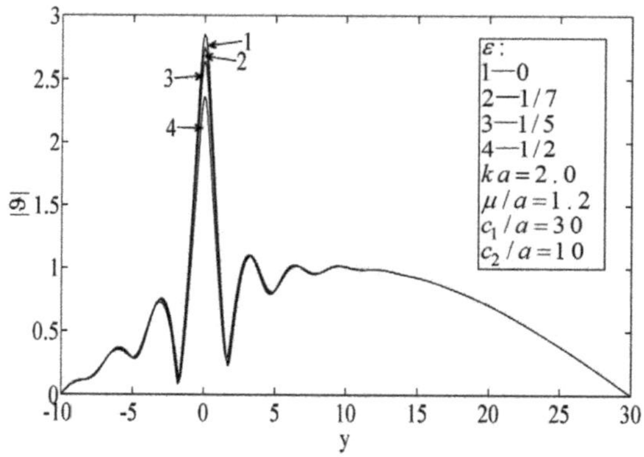

Figure 8. Temperature on the surfaces of the block.

Finally, according to Figure 9, a large wave number resulted in a high-frequency vibration (i.e., a short wave case), in which case the thermal wave propagation exhibited the properties of microparticles.

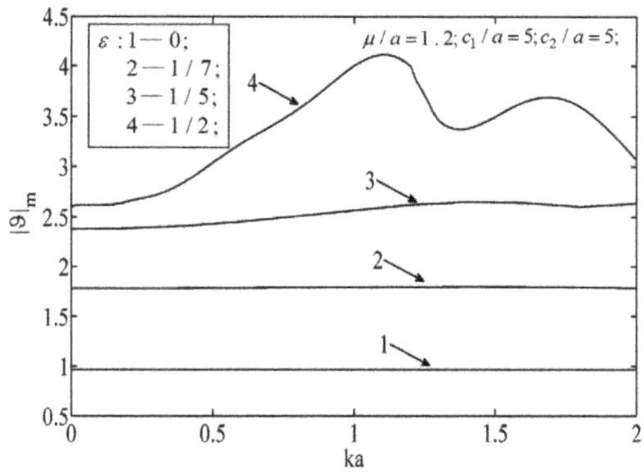

Figure 9. Temperature on the surfaces of the block.

7. Conclusions

In this paper, the propagation of thermal waves by an elliptical hole in an infinite block was investigated based on the non-Fourier law of heat conduction. A general solution of the scattering field of thermal waves based on the wave model of heat conduction was presented. The distributions and variations of temperature amplitude under different parameters were analyzed and discussed. The following conclusions can be drawn:

(1) When the wave number was comparatively large, the effects of varying temperature were great. Otherwise, the effect was narrow.

(2) The eccentricity of the ellipse had a great effect on the temperature distribution. When the eccentricity was greater, the temperature drop from the circular case was greater. When the major radius of the ellipse was along the lateral scale, the temperature showed great fluctuation, and along the longitudinal scale, the temperature showed small fluctuation.

(3) The position of the hole in the block was another factor that influenced the temperature distribution. When the center of the hole was on the median of the block, the temperature curve was symmetrical; otherwise, the curve was asymmetrical.

(4) When the scale of the block was small, the maximum temperature at the surface of the materials was great.

(5) The shallower the hole was, the greater the effect of the hole on the temperature.

(6) The thermal wave propagation characteristics showed microparticle properties under higher-frequency heat load.

Furthermore, the elliptical hole is only an example, and if proper mapping functions $\Omega(\zeta)$ are given, the application range of this method to solving the problem of the scattering of thermal waves can be spread to arbitrarily shaped holes. Of course, our above study also has some limitations. The object in this research was a block with two boundaries, and increasing the number of boundaries would mean that the results from using this theory would not be sufficiently accurate. An experimental verification system should be established in any future studies. The theory and numerical results of this paper can be used for infrared thermal wave imaging, the physical inverse problem, and the evaluation of internal holes in materials.

Author Contributions: C.Z. organized the content; C.Z. wrote Section 2, Section 3, Section 4, and Section 5; B.W. wrote Section 1, Section 6, and Section 7. All of the authors edited the final paper.

Funding: This research was funded by the National Natural Science Foundation of China (NSFC), grant number 51805124.

Conflicts of Interest: The authors declare no conflict of interest.

Appendix A

The complex function form of the scattering field of thermal waves by non-circular holes:

$$
\begin{aligned}
\vartheta^{(s)} &= \sum_{n=-\infty}^{\infty} A_n H_n^{(1)}(\kappa|z|)\left\{\tfrac{z}{|z|}\right\}^n + \sum_{n=-\infty}^{\infty} A_n \left\{ \sum_{m=1}^{\infty} (-1)^1 \left[H_n^{(1)}(\kappa|z-z_{1m}|)\left\{\tfrac{z-z_{1m}}{|z-z_{1m}|}\right\}^{(-1)n}\right] \right. \\
&+ \sum_{m=1}^{\infty} (-1)^2 \left[H_n^{(1)}(\kappa|z-z_{2m}|)\left\{\tfrac{z-z_{2m}}{|z-z_{2m}|}\right\}^{(-1)^2 n}\right] + \sum_{m=1}^{\infty} (-1)^3 \left[H_n^{(1)}(\kappa|z-z_{3m}|)\left\{\tfrac{z-z_{3m}}{|z-z_{3m}|}\right\}^{(-1)^3 n}\right] \\
&\left. + \cdots + \sum_{m=1}^{\infty} (-1)^l \left[H_n^{(1)}(\kappa|z-z_{lm}|)\left\{\tfrac{z-z_{lm}}{|z-z_{lm}|}\right\}^{(-1)^l n}\right] \right\}
\end{aligned}
$$

For the convenience of the calculation, we took $l = 4$, and the scattered field of the thermal wave field can be described as:

$$
\vartheta^{(s)} = \sum_{n=-\infty}^{\infty} A_n H_n^{(1)}(\kappa|z|)\left\{\frac{z}{|z|}\right\}^n + \sum_{n=-\infty}^{\infty} A_n \sum_{l=1}^{4} \sum_{m=1}^{\infty} (-1)^l H_n^{(1)}(\kappa|z-z_{lm}|)\left\{\frac{z-z_{lm}}{|z-z_{lm}|}\right\}^{(-1)^l n}
$$

where $z_{1m} = \mathrm{i}2(mL - c_2)$, $z_{2m} = \mathrm{i}2mL$, $z_{3m} = -\mathrm{i}2[(m-1)L + c_2]$, $z_{4m} = -2\mathrm{i}mL$; $L = c_1 + c_2$; $m = 1, 2, \cdots \infty$; $\exp(\mathrm{i}\varphi) = \Omega(\zeta)/|\Omega(\zeta)|$, $r = |\Omega(\zeta)|$, $z = \Omega(\zeta)$, $z' = \Omega'(\zeta)$.

Appendix B

The expression of E_n and E:

$$
\begin{aligned}
{}_n &= \tfrac{1}{2}\kappa \mathrm{Re}\left[\tfrac{\zeta z' \bar{z}}{|\zeta||z||}\right]\left[H_{n-1}^{(1)}(\kappa|z|) - H_{n+1}^{(1)}(\kappa|z|)\right]\left\{\tfrac{z}{|z|}\right\}^n + n\mathrm{i}\mathrm{Im}\left[\tfrac{\zeta z'}{|\zeta|z}\right] H_n^{(1)}(\kappa|z|)\left\{\tfrac{z}{|z|}\right\}^n \\
&+ \tfrac{1}{2}\kappa \sum_{m=1}^{\infty}\sum_{l=1}^{4} (-1)^l Re\left[\tfrac{\zeta z'(\bar{z}-\overline{z_{lm}})}{|\zeta||z-z_{lm}|}\right]\left[H_{n-1}^{(1)}(\kappa|z-z_{lm}|) - H_{n+1}^{(1)}(\kappa|z-z_{lm}|)\right]\left\{\tfrac{z-z_{lm}}{|z-z_{lm}|}\right\}^{(-1)^l n} \\
&+ n\mathrm{i} \sum_{m=1}^{\infty}\sum_{l=1}^{4} Im\left[\tfrac{\zeta z'}{|\zeta|(z-z_{lm})}\right] H_n^{(1)}(\kappa|z-z_{lm}|)\left\{\tfrac{z-z_{lm}}{|z-z_{lm}|}\right\}^{(-1)^l n}
\end{aligned}
$$

$$
= -\vartheta_0\{q\sin\varphi\cos[q(c_2+y)] + \mathrm{i}p\cos\varphi\sin[q(c_2+y)]\}\exp[\mathrm{i}p(x+b)]
$$

where $z = re^{\mathrm{i}\varphi}$, $z = \Omega(\zeta)$; $\zeta = \rho e^{\mathrm{i}\theta}$; $\cos\varphi = \mathrm{Re}\frac{z}{|z|}$, $\sin\varphi = \mathrm{Im}\frac{z}{|z|}$; $X_n = A_n$.

References

1. Thomas, R.L.; Favro, L.D.; Grice, K.R. Thermal wave imaging for nondestructive evaluation. *Can. J. Phys.* **1982**, *64*, 586–590.
2. Qiu, T.Q.; Tien, C.L. Short-pulse laser heating on metals. *Int. J. Heat Mass Transf.* **1992**, *35*, 719–726. [CrossRef]
3. Maillard, S.; Cadith, J.; Bouteille, P. Non-destructive testing of forged metallic materials by active infrared thermography. *Int. J. Thermophys.* **2012**, *33*, 1982–1988. [CrossRef]
4. Ping, S. *Mesoscopic Phenomena—Introduction to Wave Scattering, Localization, and Mesoscopic Phenomena-10*; Springer: Berlin/Heidelberg, Germany, 2006.
5. Fang, X.Q.; Liu, J.X.; Hu, C. Nondestructive evaluation of a conducting sphere in semi-infinite functionally graded materials using thermal wave method. *NDT E Int.* **2010**, *43*, 163–170. [CrossRef]
6. Garrido, F.; Salazar, A. Thermal wave scattering by spheres. *J. Appl. Phys.* **2004**, *95*, 140–149. [CrossRef]
7. Tzou, D.Y. A unified field approach for heat conduction from macro-to-micro-scales. *J. Heat Transf.* **1995**, *117*, 8–16. [CrossRef]
8. Tang, D.W.; Araki, N. Non-fourier heat conduction in a finite medium under periodic surface thermal disturbance. *Int. J. Heat Mass Transf.* **1996**, *39*, 1585–1590. [CrossRef]
9. Ma, X.B.; Ye, S.L. Scattering of thermal waves by a heterogeneous subsurface spheroid inclusion including non-fourier effects. *Thermochim. Acta* **2014**, *597*, 65–70. [CrossRef]
10. Bouteille, P.; Legros, G.; Bodnar, J.L. Non-destructive testing of metallic materials using passive, and active infrared thermography. *Mech. Ind.* **2014**, *15*, 313–321. [CrossRef]

11. Thibaud, J.B.; Carminati, R.; Greffet, J.J. Scattering of a diffusive wave by a subsurface object. *J. Appl. Phys.* **2000**, *87*, 7638–7646. [CrossRef]
12. Amirian, E.; Dejam, M.; Chen, Z. Performance forecasting for polymer flooding in heavy oil reservoirs. *Fuel* **2018**, *216*, 83–100. [CrossRef]
13. Saboorian-Jooybari, H.; Dejam, M.; Chen, Z. Heavy oil polymer flooding from laboratory core floods to pilot tests and field applications: Half-century studies. *J. Pet. Sci. Eng.* **2016**, *142*, 85–100. [CrossRef]
14. Salazar, A.; Sánchez-Lavega, A.; Celorrio, R. Scattering of cylindrical thermal waves in fiber composites: In-plane thermal diffusivity. *J. Appl. Phys.* **2003**, *93*, 4536–4542. [CrossRef]
15. Terron, J.M.; Sanchez-Lavega, A.; Salazar, A. Multiple scattering effects of thermal waves by two subsurface cylinders. *J. Appl. Phys.* **2000**, *87*, 2600–2607. [CrossRef]
16. Wang, F.; Ma, X.B.; Chen, D.Z. Thermal wave scattering in functionally graded materials containing a spherical inclusion. *Thermochim. Acta* **2015**, *600*, 116–123. [CrossRef]
17. Terron, J.M.; Sanchez-Lavega, A.; Salazar, A. Multiple scattering of thermal waves by a coated subsurface cylindrical inclusion. *J. Appl. Phys.* **2001**, *89*, 5696–5702. [CrossRef]
18. Ma, X.B.; Jiang, H.Q.; Chen, D.Z. Thermal wave scattering by a cylindrical subsurface inclusion in semi-infinite slab. *Appl. Therm. Eng.* **2016**, *100*, 691–698. [CrossRef]
19. Muskhelishvili, N.I. *Some Basic Problems of the Mathematical Theory of Elasticity*; International Publishing: Noordhoff, The Netherlands, 1956.

sensors

Review

PEDOT:PSS-Based Conductive Textiles and Their Applications

Granch Berhe Tseghai [1,2,3,*], Desalegn Alemu Mengistie [2,4], Benny Malengier [1], Kinde Anlay Fante [3] and Lieva Van Langenhove [1]

1 Department of Materials, Textiles and Chemical Engineering, Ghent University, 9000 Gent, Belgium; Benny.Malengier@UGent.be (B.M.); Lieva.VanLangenhove@UGent.be (L.V.L.)
2 Ethiopian Institute of Textile and Fashion Technology, Bahir Dar University, 6000 Bahir Dar, Ethiopia; dmengist@calpoly.edu
3 Jimma Institute of Technology, Jimma University, Jimma, Ethiopia; kinde.anlay@ju.edu.et
4 Materials Engineering Department, California Polytechnic State University, San Luis Obispo, CA 93407, USA
* Correspondence: GranchBerhe.Tseghai@UGent.be; Tel.: +32465570635

Received: 28 February 2020; Accepted: 26 March 2020; Published: 28 March 2020

Abstract: The conductive polymer complex poly (3,4-ethylene dioxythiophene):polystyrene sulfonate (PEDOT:PSS) is the most explored conductive polymer for conductive textiles applications. Since PEDOT:PSS is readily available in water dispersion form, it is convenient for roll-to-roll processing which is compatible with the current textile processing applications. In this work, we have made a comprehensive review on the PEDOT:PSS-based conductive textiles, methods of application onto textiles and their applications. The conductivity of PEDOT:PSS can be enhanced by several orders of magnitude using processing agents. However, neat PEDOT:PSS lacks flexibility and strechability for wearable electronics applications. One way to improve the mechanical flexibility of conductive polymers is making a composite with commodity polymers such as polyurethane which have high flexibility and stretchability. The conductive polymer composites also increase attachment of the conductive polymer to the textile, thereby increasing durability to washing and mechanical actions. Pure PEDOT:PSS conductive fibers have been produced by solution spinning or electrospinning methods. Application of PEDOT:PSS can be carried out by polymerization of the monomer on the fabric, coating/dyeing and printing methods. PEDOT:PSS-based conductive textiles have been used for the development of sensors, actuators, antenna, interconnections, energy harvesting, and storage devices. In this review, the application methods of PEDOT:SS-based conductive polymers in/on to a textile substrate structure and their application thereof are discussed.

Keywords: PEDOT:PSS; wearable electronics; e-textile; conductive textile

1. Introduction

With the emergence of new fibers, fabrics and innovative processing technologies, the growth of the textile market has increased in recent years and has been instrumental in bringing about significant technological advances. Starting with groundbreaking research on how to integrate conductive lines and circuits into textiles in the late 1990s, rigorous researches resulted in sensor additions, actuators, user interfaces, and complicated textile circuits that could provide extra functionality to make smart textiles. Smart textiles can be defined as textiles capable of sensing and responding to changes in their environment by external factors. In response to stimuli, they are able to show significant changes in their mechanical and/or chemical properties (such as shape, color, and stiffness), or in their thermal, optical, or electromagnetic properties. Examples include fabrics that change their color with changes in temperature and fabrics that regulate garments' surface temperature to achieve physiological comfort. Smart materials can be incorporated into the textile structure by different technologies such as embroidering [1], non-woven textile [2], knitting [3], weaving [4], braiding [5], yarn spinning [6], fiber spinning [7], polymerizing [8], coating [9], plating [10] and printing [11].

Societal needs such as new functionality, comfort, and aesthetic values from daily use to critical health-related applications are the driving forces for the development of smart textile materials. The recent developments in the fields of textiles, electronics, information technology, advanced materials and polymers are paving the way for the development of smart textile materials and their application [12–19]. The fact that textiles are an interface between the wearer and the surrounding with large and permanent surface contact make them ideally suitable for large scale and long-term health monitoring. In addition, textile is easily accessible anywhere and has versatile applications from dressing to household products and coverings.

The primary step in smart textiles is making conductive textiles. From a textile perspective, it is suggested that the overall objective of smart textiles would be to convert all related components, such as sensors, actuators, transmission lines, etc., into 100% textile materials [20]. However, conventional textiles are intrinsically non-conductive so they need to be converted to become conductive in some way. The earlier method of making textiles conductive was inserting thin metal fibers/yarns in the textile/garment which primarily was developed for antistatic treatment. A later development and more convenient way to make conductive textiles is to treat them with conductive inks at the polymer (man-made fibers), fiber, yarn, fabric or readymade garment stage. Conductive textiles can be classified into bulk and surface conductive textiles [21]. Bulk conductive textiles include intrinsically conductive polymer textiles, textiles twisted/embedded with metallic filaments and textiles filled with conductive additives such as carbon blacks (CBs), carbon nano-tubes (CNTs), or conductive polymers. Surface conductive textiles are the textiles coated with conductive layers. The conductive coatings may consist of metals, conductive polymers, or other conductive materials such as CNTs or CBs. From a technological point of view, the challenge is to develop a method suitable for the current textile processing, with high conductivity, durability and maintaining the desired textile properties such as flexibility. Coating with metal nanoparticles/nanowires could give high conductivity; however, it may come at the expense of flexibility and lack of durability. The focus will therefore be on enhancing durability, textile character, and conductivity. The use of electrically conductive polymeric materials have recently attracted considerable interest from academic and industrial researchers to explore their potential in sensors [22], biomedical [23], wireless communication patch antenna [24], energy harvesting [25] and energy storage [26] applications. Conductive polymers are light weight and flexible and can be applied on the textile without affecting its flexibility. Solution-based conductive polymers are especially convenient for the roll-to-roll processing which can easily be integrated with the current textile processing technologies like dyeing and printing. There are several reports [27–29] on the use of conductive polymers for conductive textile for different applications. Recently reported conductivities of over 6000 S/cm [30] are signaling their practical potential use in the smart textiles applications.

For this review, we made a comprehensive electronic document search according to the preferred reporting items for Systematic Reviews and Meta-Analyses (PRISMA) guidelines from the web of science database and the Google search engine. 'Conductive + textiles', 'textile + sensor', 'textile + antenna', 'textile + energy harvesting', 'textile + energy storage', 'textile + interconnections', 'conductive + polymers', 'conductive + polymer + composites', 'e-textile', 'PEDOT', 'PEDOT:PSS', 'EDOT', 'poly(3,4-ethylenedioxythiophene)', 'poly(3,4-ethylene dioxythiophene):polystyrene sulfonate', have all been used as primal keywords for the search.

2. Conductive Polymers

Traditional commodity polymers are intrinsically insulators. The discovery of conductive polymers started with the path breaking discovery that halogen doped polyacetylene $(\text{-CH=CH-})_n$ show high electrical conductivity, which led to the 2000 Nobel Prize in Chemistry award [31]. Since then, there have been several fundamental studies and applications of conductive polymers. The carbon atom in saturated polymers, such as polyethylene, form four covalent σ-bonds (saturated sp^3-carbon).

Whereas the carbon atom in conjugated polymers has sp^2p_z (π) orbitals which form three σ-bonds and the remaining p_z orbitals engage in the π system. The common feature in conductive polymers is conjugation, i.e., the alternation of single and double bonds, and hence the synthesis of π-conjugated chains is central to the science and technology of conductive polymers. The charge carriers are delocalized in conjugated systems and provide the "highway" for charge mobility along the backbone of the polymer chain. The conductivities of conjugated polymers can be enhanced by doping, which is basically either reduction or oxidation [32]. The conductivity of doped polyacetylene can reach 10^5 S/cm which is comparable to that of copper [33]. However, polyacetylene is difficult to synthesize and is unstable in air which prevented its commercialization. The most important conductive polymer candidates currently are polypyrrole (PPy), polyaniline (PANI), and polythiophenes (PTh) whose chemical structures are shown in Figure 1. Poly(3,4-ethylenedioxythiophene) (PEDOT), which is the main topic of this article, is the most studied and successful PTh derivative polymer due to its higher electrical conductivity and chemical stability which make it suitable in the development of smart textiles [34]. In contrast to PPy and PANI, the exploration on PEDOT is comparatively recent.

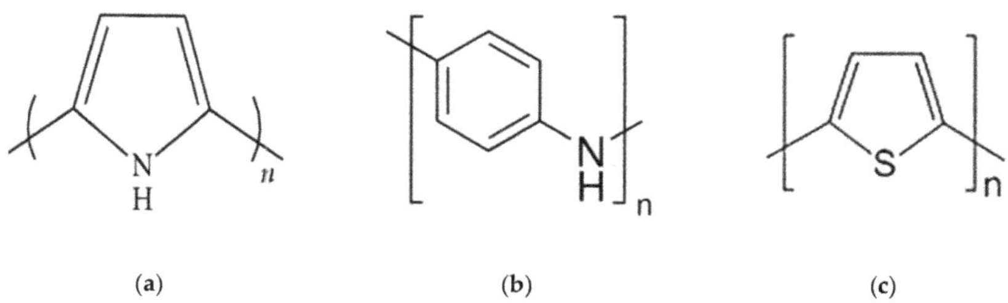

Figure 1. Chemical structure of the most common conductive polymers: (**a**) Polypyrrole; (**b**) Polyaniline; (**c**) Polythiophene.

Conductive polymers exhibit novel properties such as solution processability, high elasticity, toughness, and low-temperature synthetic routes. Some examples conductive polymers and their properties are presented in Table 1. Due to these interesting properties, conductive polymers are used for several applications such as photovoltaic devices [35], organic light-emitting diodes [36], organic field-effect transistors [37], sensors [38], antennas [39], conductive textiles [40], supercapacitors [41] and many more.

Table 1. Non-exhaustive conductivity and properties of common conductive polymers.

Polymer	Conductivity (S/cm)	Doping	Properties	Limitations	Ref.
PPy	2000	P	High electrical conductivity, ease of preparation and ease of surface modification	Rigid, brittle and insoluble	[42]
PANI	112	P	Diverse structural forms, environmentally stable, low cost	Hard to process, non-biodegradable, limited solubility	[43]
PTh	560	P	High electrical conductivity, ease of preparation, good optical property	Hard to process	[29]
PEDOT:PSS	4700	P	High electrical conductivity used as transparent electrode	Needs additional steps to process	[44]

3. PEDOT

Among conductive polymers, PEDOT is the most extensively explored, successful and widely used for many applications due to its high conductivity, its stability in air up to high temperatures and resistance to humidity including moist air, and because it is also processable in water. PEDOT can be polymerized from 3,4-ethylenedioxythiophene (EDOT) chemically or electrochemically. However, PEDOT synthesized this way and doped with small molecule counter ions is insoluble in any solvent and large size sample preparations are a challenge [32].When polymerization is carried out in the presence of aqueous polyelectrolyte poly(styrenesulfonate) (PSS), it becomes water dispersible which is stable, easy to process, with good film forming properties, and with high visible light transmittance. PSS acts as a template during polymerization and charge balancing counter ion hence keeping the cationic PEDOT segments dispersed in aqueous medium. The molecular weight of PEDOT and PSS is about 1000–2500 g/mol (around 10 to 20 monomer units) and 400,000 g/mol, respectively. PEDOT:PSS in the aqueous media (and the as-prepared film too) has core-shell structure (Figure 2) where the core is conductive PEDOT-rich and the shell is insulator PSS-rich. The hydrophobic PEDOT and hydrophilic PSS nature led to the core shell structure [45]. PEDOT:PSS films prepared from aqueous dispersions have lower conductivity (<1 S/cm) than PEDOT films prepared by oxidative and vapor phase polymerization and stabilized with small molecule counter ions. The main reason for the low conductivity is the core-shell structure which leads to an increase in the energy barrier for charge transport across PEDOT chains by the insulator PSS-rich shell and charge localization due to the coiled PEDOT-rich core [46].

The conductivity can be enhanced up to four orders of magnitude by treatment with polar solvents like dimethyl sulfoxide, ethylene glycol, acids and alcohols called "secondary dopants". Secondary dopants are different from primary dopants in that they are apparently "inert" and the newly enhanced property persists even upon complete removal of the secondary dopants. Generally, the treatment methods can be grouped into three types: mixing secondary dopant in to the aqueous PEDOT:PSS dispersion, film treatment after drying with secondary dopant or a combination of both methods. The exact mechanism of conductivity enhancement is still a topic of intense investigation. Shi et al. have nicely reviewed treatment methods for conductivity enhancement and mechanism of conductivity enhancement [28]. The additives bring about charge screening between PEDOT and PSS due to their high dielectric constant leading to phase separation. The PEDOT chains will be free to be linearly oriented (from coiled structure), and hence, have a more compact structure (smaller π–π stacking distance), leading to stronger inter chain coupling and better crystallinity with larger crystal size [47]. In the case of post treatment, the excess PSS will also be removed [48]. All these combined effects will lead to increases in carrier concentration and mobility [49,50].

There are different grades of PEDOT:PSS commercially available with different conductivities, may be due to the molecular weight difference of PEDOT. Recently, the most extensively used high conductivity grade is Clevios PH1000. Rigorous work has shown very high conductivities of 4700 S/cm for PEDOT:PSS [44] and 7619 S/cm for single crystal PEDOT nanowires [51]. With such improved conductivities, further advancements in different applications are expected.

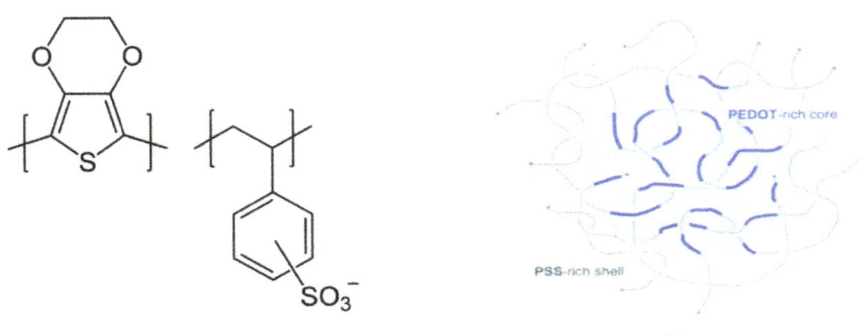

Figure 2. (**a**) Chemical structure of PEDOT:PSS; (**b**) core-shell structure of PEDOT:PSS, adapted from [52].

4. PEDOT:PSS Based Conductive Polymer Composites

4.1. Conductive Polymer Composites

Metal-based interconnects have been reported to have the highest conductivity, but are not stretchable enough, while elastomeric interconnects are not conductive enough. Conventional conductive polymers such as PPy and PEDOT show promising conductivity for these applications; however, their mechanical properties, biocompatibility and processability still needs improvement [34]. This has led to more attention being directed towards conductive polymeric composites at improved electrical conductivity and mechanical stability. One way to increase the mechanical robustness of conductive polymers is by making a composite with commodity polymers. Composite materials based on conjugated conducting polymers and non-conducting polymers often show a low percolation threshold and improved environmental stability with respect to the conjugated polymer. For instance, compounding techniques used for processing of conventional thermoplastics have been applied to prepare composites of PPy with certain thermoplastics which provided a drastic increase in oxidation stability [53]. In particular, composite of conductive polymer with elastomers have been demonstrated for stretchable/elastic conductive materials/devices. Typical examples of conductive polymer composites for different applications include an electrically conductive PEDOT:PSS-polyurethane (PU) [54]. Table 2 presents non-exhaustive lists of common conductive polymer composites with their suggested application areas.

Table 2. Properties of common conductive polymer composites.

Conductive Polymer Composite	Conductivity (S/cm)	Properties	Suggested Application	Ref.
PPy/Hyaluronic Acid	3.1×10^{-3}	Can support tissue growth and stimulate specific cell functions	Tissue engineering and wound-healing	[55]
PANI Nanofibers/Collagen	0.27	Well suited for cell culture	Scaffold Materials for biomedical	[56]
PPy/Chitosan	10^{-3}–10^{-7}	Radical scavenger	Food packaging and biomedical	[57]
PEDOT:Tos/Glycol	1486	Soft, flexible and biocompatible	Implantable devices	[58]
PPy/Cellulose Acetate	6.9×10^{-4}–360	Soft and flexible and	Wearable electronics	[59]
PANI Nanoparticles/Polyacrylic Acid/Polyvinyl Alcohol	0.04–0.06	Hydrogel, biocompatible, good mechanical strength and good swelling properties	Strain sensor	[60]
Polythiophene derivative/PU	2.2×10^{-5}	Suitable for supporting electrically stimulated cell growth	Tissue engineering	[23]
PEDOT:PSS/PU/Ionic liquid	8.8×10^{-5}	Mechanically flexible and stretchable	Actuating devices	[61]
PPy/poly(D,L-Lactic Acid)	5.7×10^{-3}–15.7×10^{-3}	Nerve tissue regeneration, biocompatibility	Synthetic nerve conduits	[62]
PPy nanoparticles/PU	2.3×10^{-6}	Cytocompatible, elastomeric properties	Tissue engineering	[63]

4.2. PEDOT:PSS Based Conductive Polymer Composites

PEDOT:PSS is well known for its high conductivity and applications in conductive textiles and has been used with encouraging results for different applications. Unfortunately, the use of pure PEDOT:PSS is currently constrained by its brittleness. As outlined earlier, one way to improve its mechanical flexibility is to make a composite with traditional commodity polymers. Giuri et al. reported are GGO-PEDOT composites with thermal stability up to 270 °C for super capacitors [41], Hilal and Han developed a graphene (G) and PEDOT:PSS composites with improved electrical conductivity by 63% of a pristine PEDOT:PSS for solar cells [64]. Taroni et al. reported a thermoelectric PEDOT:PSS/PU blend [38] with improved ductility while maintain reasonable conductivity. A polyvinyl alcohol (PVA) combined with phosphoric acid and PEDOT:PSS and silver flakes that withstands about 230% strain before fracture was reported by Houghton et al. [65]. Furthermore, a PEDOT:PSS-based multi-layer bacterial composite was developed by embedding an electro-active bacterium inside a conductive three-dimensional PEDOT:PSS matrix to increase the electron transfer through the PEDOT:PSS [66]. Table 3 presents non-exhaustive lists of PEDOT:PSS composites with their preparation technique, properties and proposed applications.

Table 3. Properties of common PEDOT:PSS based conductive polymer composites.

Conductive Polymer Composite	Resistivity, Ω cm (Resistance, Ω/sq)	Properties	Manufacturing Technique	Proposed Application	Ref.
GO/rGO filled PVA/PEDOT:PSS	10^7	Highly flexible free-standing	Solvent casting	Strain sensor	[67]
PEDOT:PSS/PU	(35–240)	Highly flexible, stretchable	Electrospinning	Strain sensor	[2]
PEDOT:PSS/Bacteria	103 [a]	20 times more steady-state current than native biofilms baseline with signal level of 6.31 μA/cm^3	Embedding bacteria into electro-polymerized PEDOT:PSS on carbon felt anodes	Bioelectronics	[66]
PEDOT:PSS/PU	1.26×10^{-2}	Sensitivity to different stimuli including strain, ambient temperature and/or air flow high electrical conductivity	Dispersion mixing	Stretchable self-powered sensors	[38]
PEDOT:PSS/graphene	(25)	Strong resistance against fatigue upon repeated folding-unfolding	Spray coating	Data storage and transmission, biosensors and actuators	[68]
3D Graphene/PEDOT:PSS	4.1×10^{-2}	Good resistance retention capability under deformations	Graphene networks coated by PEDOT:PSS	Next-generation stretchable electronics	[69]
CNT/PEDOT:PSS	3×10^{-3}	Good thermoelectric performance	Vacuum assisted filtration method and H_2SO_4 treatment	Flexible thermoelectric generator	[70]

[a] The unit for the value of Ref. [66] is Ω.

5. Methods of Treating Textiles with PEDOT:PSS

PEDOT:PSS can be applied on textile materials by carrying out an in-situ polymerization of 3,4-ethylenedioxythiophene (EDOT) on the textile substrate in the presence of PPS or by applying the polymer PEDOT:PSS dispersion onto a textile substrate. In general, adding the polymer into a polymer solution during fiber spinning, coating/dyeing textile substrates (fibers, yarns, fabrics) and/or printing textile fabrics, can be used to produce PEDOT based conductive textiles.

5.1. Conductive Fiber Spinning

In this technique, PEDOT:PSS is added to a conventional polymer solution during fiber wet spinning or electrospinning (Figure 3a) in order to produce a conductive fiber or filament, or the PEDOT:PSS alone can be spun in to a fiber. In 2003, Okuzaki and Ishihara presented their first study on the manufacture of 4.6 to 16 μm PEDOT:PSS microfibers using wet-spinning technique with an electrical conductivity of 0.1 S/cm [7]. The Young's modulus, tensile strength, and elongation at break for the resulting microfibers were 1.1 GPa, 17.2 MPa, and 4.3%, respectively. Jalili et al. reported a simplified wet-spinning process for continuous PEDOT:PSS fibers which showed a conductivity up to 223 S/cm by post treatment of the fibers with ethylene glycol [71]. In another approach, they used an aqueous blend of PEDOT:PSS and poly(ethlylene glycol) and the conductivity of the fibers increased by a 30-fold (from 9 to 264 S/cm) without the need of a post treatment. Okuzaki et al. developed PEDOT:PSS microfibers with diameter of ca. 5 μm by wet-spinning [72]. They improved the electrical conductivity of the fibers from 74 S/cm to 467 S/cm by subsequent dip-treatment of the fibers in ethylene glycol. The mechanical properties of the microfibers were also improved by the dip-treatment; the Young's modulus and tensile strength increased from 3.2 GPa and 94 MPa to 4.0 GPa and 130 MPa, respectively. Zhou et al. further enhanced the electrical conductivity of wet spun PEDOT:PSS microfibers to 2804 S/cm via wet-spinning followed by post treatment with ethylene glycol and hot-drawing [73]. This high conductivity is due to the combined effects of the vertical hot-drawing process and doping/de-doping of the microfibers with ethylene glycol. Moreover, they had a semiconductor metal transition at 313 K with superior mechanical properties with a Young's modulus up to 8.3 GPa, a tensile strength reaching of 409.8 MPa and a large elongation before failure (21%). J. Zhang et al. also carried out a wet spinning of PEDOT:PSS fiber and obtained better conductivity of PEDOT:PSS fiber, 3828 S/cm, by decreasing the fiber diameter using a fine gauge needle [74]. The wet-spinning set-up was modified as shown in Figure 4a. Liu et al. also reported composite conductive fibers based on PEDOT:PSS blended with polyacrylonitrile [75] by wet spinning. Fibers with 1.83 wt% of PEDOT:PSS showed a conductivity of 5.0 S/cm. Seyedin et al. demonstrated a scaled-up fiber wet-spinning production of electrically conductive and highly stretchable PU/PEDOT:PSS fibers which were then used in knitting for a knee sleeve prototype with application in personal training and rehabilitation following injury [3]. The fiber showed a conductivity of 166 S/cm very close to pristine PEDOT:PSS film with a wide strain sensing capability up to a 260 % strain.

Jin et al. employed an electrospinning and in-situ synthesis process to fabricate silver nanoparticles coated PEDOT:PSS/PVA flexible self-supporting nanofibers with greatly improved electrical conductivity [76]. Q. Zhang et al. also used an electrospinning to fabricate a PVA/PEDOT:PSS nanofiber with an average diameter of 68 nm for a gas sensor (Figure 3b) [77].

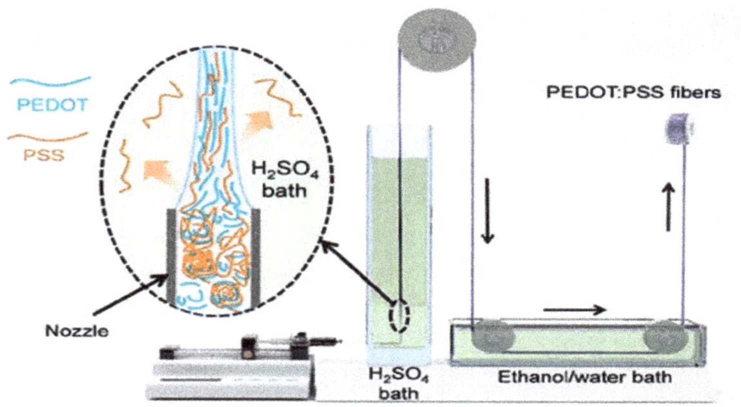

(a)

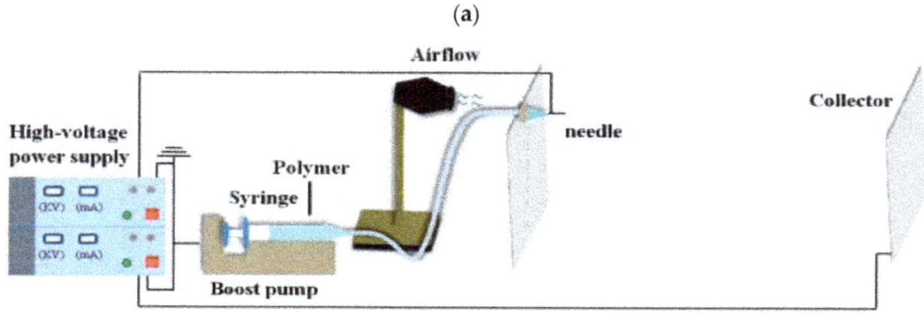

(b)

Figure 3. Fiber spinning process: (**a**) schematic illustration modified set-up used in wet-spinning PEDOT:PSS fibers. Inset shows the schematic illustration of the alignment of PEDOT:PSS chains during fiber formation and the outward diffusion of excess PSS to H_2SO_4 coagulation bath. Adapted from [74]; (**b**) schematic diagram of electrospinning setup; the distance between the needle and collector was 120 cm. Adapted from [77].

5.2. Polymerization of PEDOT on the Textile Substrate

The PEDOT monomer can be polymerized on the textile substrate (fiber, yarn, fabric or garment form) by in situ, vapor phase or electrochemical polymerization by using EDOT and appropriate chemicals like oxidants [78]. This method combines polymerization of the PEDOT and coating of the textile.

The attachment of PEDOT on the fabric surface depends on the chemistry of the fiber as well as the surface roughness of the fiber. Though direct polymerization of PEDOT on the textile seems straight forward, it is difficult to control the parameters. Moreover, it is used for small sample size and a challenge for industrial requirements. Hong et al. carried out five cycles of in-situ polymerization of PEDOT on poly(trimethylene terephthalate) fabrics in the presence of ferric p-toluenesulfonic acid and ferric chloride as oxidants followed by butane treatment and obtained an electrical conductivity of 3.6 S/cm [79]. Bashir et al. reported an electrically conductive polyester fabric with an electrical resistance of ~2000 Ω, coated by PEDOT through oxidative vapor phase polymerization (VPP) in the presence of Fe (III) chloride hexahydrate oxidant [80]. They also obtained electro-conductive aramid, viscose and nylon fabrics by the same approach. In another work, they produced a conductive viscose yarn with electrical resistance 6 kΩ by oxidative chemical vapor deposition, by removing the impurities like acetone and ethyl acetate, prior to the oxidant enrichment and polymerization steps [81]. Trindade et al. also coated a polyester fabric by PEDOT through VPP and obtained a lower sheet resistance of~20 Ω/sq by increasing the concentration of the oxidant, Fe (III) chloride hexahydrate [82]. L. Zhang et al. coated textile fabrics (silk, linen, wool, pineapple, bamboo rayon) by PEDOT through VPP and obtained a sheet resistance from 200 to 9.46 kΩ depending on the porosity of the fabric; porous fabric gives higher sheet resistance than tight-fabric [83]. Overall, the electrical and mechanical properties of conductive textiles are determined by the concentration of oxidants, pretreatment of the original pristine fabric and post-treatments of the conductive fabric, type and form of textile substrate and the polymerization conditions. The illustration of vapor deposition system is shown in Figure 4.

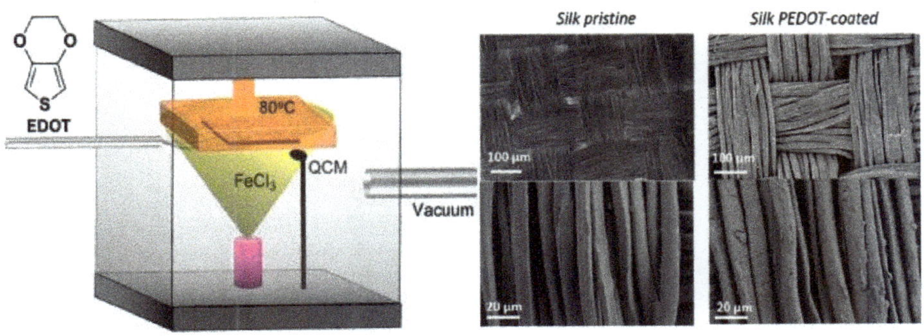

(a) (b)

Figure 4. (**a**) The schematic illustration of vapor deposition system for PEDOT; (**b**) SEM images of pristine silk textile and PEDOT-coated silk textile. Adapted from [83].

5.3. Coating/Dyeing of Textiles with PEDOT:PSS

In the coating/dyeing method, the appropriate form of textile is treated by immersing/dipping in PEDOT:PSS dispersion with appropriate auxiliary chemicals. This is method mimics either the exhaust or continuous dyeing method of commercial textile processing. It is the most popular method practiced for making conductive textiles with PEDOT:PSS. The uniformity as well depth of dyeing/coating depends on the functional group of the textile. Ding et al. treated cotton, cotton/polyester, polyester and nylon/spandex fabrics by impregnating with PEDOT:PSS and showed that conductivity is higher for fabrics which swell well in water [84]. Ryan et al. dyed up to 40 m long silk yarn with PEDOT:PSS with conductivity of ~14 S/cm which was durable to machine washing [85]. The reason to wash durability of PEDOT:PSS on silk is due to the dyeing effect and the presence of a fluorosurfactant Zonyl FS-300 used during dyeing. When cotton was dyed by the same method, it was too fragile due to hydrolysis of the cellulose by the strong acidic PEDOT:PSS. The same group further demonstrated a continuous dyeing process to produce more than 100 m of silk thread dyed with PEDOT:PSS for a wash and wear resistant functional thread with a conductivity of about 70 S/cm [86]. Ding et al. produced PU fibrous nonwoven and treated it with PEDOT:PSS by dip-coating [2]. The PEDOT:PSS@PU nonwovens showed sheet resistance of 35–240 Ω/sq (electrical conductivity of 30–200 S/m) by varying the number of dip-coating times. This conductive nonwoven maintained its surface resistance up to 50% strain, promising for wearable application. Tadesse et al. also treated polyamide/lycra elastic fabric with PEDOT:PSS by dipping only once and showed a sheet resistance of ~ 1.7 Ω/sq [87]. The fabric was stretchable up to ~650% and maintained reasonable conductivity up to washing cycles. The durability to washing in this case is also due to dyeing effect where there is some kind of chemical interaction between the fiber and PEDOT:PSS. A schematic representation of discontinuous and continuous PEDOT:PSS dip-coating/dyeing on a textile fabric are shown in Figure 5.

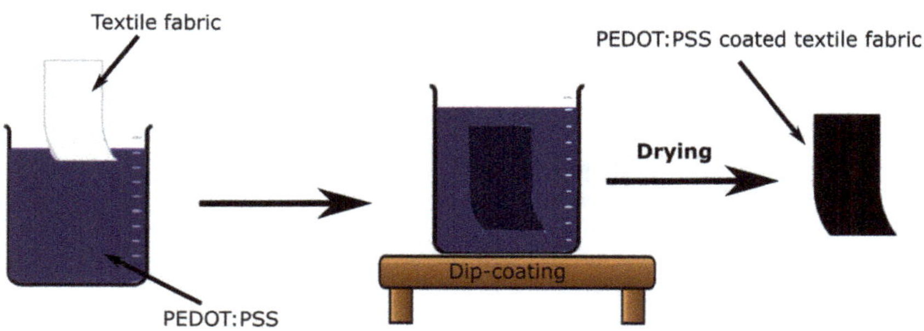

(a)

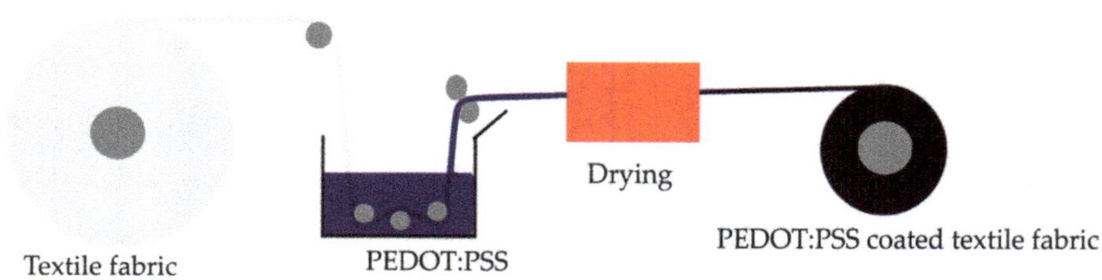

(b)

Figure 5. Schematic representation of the setup for dyeing: (**a**) discontinuous process; (**b**) continuous process.

5.4. Printing of PEDOT:PSS on Textile

Printing is a well-developed textile processing method used industrially is also used to apply the PEDOT:PSS to the textile structure in the presence of thickening agents to obtain an adequate paste or ink viscosity. Guo et al. demonstrated a fabrication of all-organic conductive wires by utilizing patterning techniques such as inkjet printing and sponge stencil to apply PEDOT:PSS onto a nonwoven polyethylene terephthalate (PET) providing a wide range of resistance, i.e., tens of kΩ/sq to less than 2 Ω/sq that allows the resistance to be tailored to a specific application [88]. Sinha et al. demonstrated the integration of screen-printed PEDOT:PSS electrocardiography (ECG) circuitry on finished textiles and recorded an ECG signal comparable to Ag/AgCl connected to copper wires [89]. Zhao et al. also used screen-printing to produce a PEDOT:PSS and carbon-based disposable electrochemical sensor for sensitive and selective determination of carmine [90]. Tseghai et al. used a flat screen printing to coat a PEDOT:PSS conductive polymer composite on to a knitted cotton fabric and obtained a sheet resistance of 24.8 kΩ/sq [91]. The conductive textile fabric stays conductive until its infliction point of stretching. The schematic illustration of screen printing is shown in Figure 6.

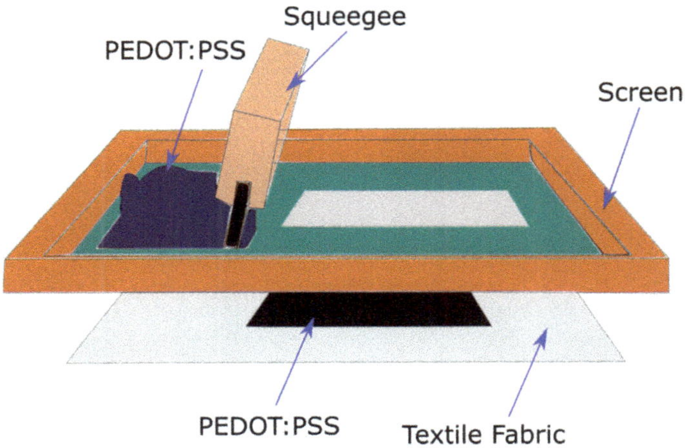

Figure 6. Schematic illustration of PEDOT:PSS screen printing on a textile fabric. Adopted from [91].

6. Applications of PEDOT:PSS-based Conductive Textiles

As outlined earlier, PEDOT:PSS has high electrical conductivity, thermal stability, decent biocompatibility, and is solution processable. These interesting properties make it attractive for different textile-based applications including sensors, energy harvesting, and storage devices.

6.1. Sensors

The demand of textile-based sensors is increasing because of their lightweight, flexiblity, and possibility of washing. PEDOT:PSS-based textiles have been widely used as a sensing component for strain, pH, humidity, biopotential, and temperature. Zahid et al. applied graphene nanoplatelets dispersed in PEDOT:PSS solutions for producing a conductive, breathable and lightweight mercerized cotton fabrics by spray coating which showed a highly repeatable and stable response to cyclic deformation tests at 5% and 10% strain rates for up to 1000 cycles with ~90% viscoelastic recovery levels after cessation [68]. Kang reported a resistive memory graphane-PEDOT:PSS coated nylon thread with a strain response for wearable applications as an example of bio-potential sensors (Figure 7) [92]. Seyedin et al. developed a strain sensor from a PU/PEDOT:PSS fibers with conductivity of 9.4 S/cm [3]. The resistance of this textile sensor stays stable up to 160% strain and up to 500 cycles. The high conductive textile-based hybrid showed high stability during stretching. Pani et al. developed a new textile ECG electrode based on woven fabrics treated with PEDOT:PSS for bio-potential recordings tested on human, both in terms of skin contact impedance and quality of ECG signals recorded at rest and during physical activity [93]. The electrode was found to be capable of operating under both wet and dry conditions, which could be an important milestone in wearable monitoring of heart. Ankhili et al. developed an ECG sensor electrode from a PEDOT:PSS screen-printed cotton fabric and obtained a clear ECG wave amplitudes up to 50 washing cycles [94]. The same group also produced washable screen-printed cotton textile electrodes with and without lycra of different PEDOT:PSS concentration, providing a medical quality ECG signal to be used for long-term ECG measurements with a similar result to silver-plated cotton fabric at 12.8 wt% of PEDOT:PSS to pure cotton [95]. Niijima et al. produced "hitoeCap" from PEDOT:PSS textile electrodes for securing electromyography of the masticating muscles [96].

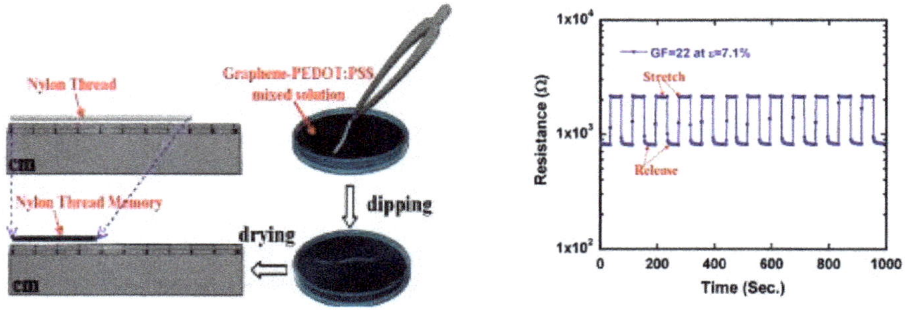

(a) (b)

Figure 7. A graphene-PEDOT:PSS coated nylon thread: (a) schematic of the simple two-step dip-and-dry solution process for the fabrication (right) and the actual picture of the sample with the length reduction from 80 to 29.48 mm after dip-and-dry; (b) resistive memory strain sensor thread at many stretch and release cycle under applied a fixed ε = 7.1%. Adopted from [92]

Furthermore, Abbasi et al. worked on the use of PEDOT:PSS material for the implementation of high sensitivity moisture sensor devices, which showed significant frequency shifts [97]. They demonstrated sensing capacity even for small moisture variations. Smith et al. developed a wearable pH sensor cotton yarn in PEDOT:PSS and multi-walled carbon nanotubes followed by PANI deposition that produced electrodes with significant biocompatibility and antibacterial properties that could be manufactured (alongside quasi-reference electrodes) into wearable solid-state pH sensors and achieved wearable pH sensors [98].

6.2. Energy Harvesting and Storage

Textiles coated with PEDOT:PSS have been used to manufacture flexible and lightweight energy harvesting and storage devices. This is quite interesting to power from wearable electronics to medical implantable devices. PEDOT:PSS is a promising organic thermoelectric material, materials which change temperature difference into electricity or vice versa [52]. PEDOT:PSS coated textiles have been studied for wearable thermoelectric applications to harvest the temperature difference between the body and outer surrounding. Du et al. coated polyester fabric strips with PEDOT:PSS where the flexibility and air permeability was not affected and attached them on non-coated [99]. The treated fabric showed electrical conductivity of ~1.5 S/cm and generated thermoelectric voltage of 4.3 mV at a temperature difference of 75.2 °K. Ryan et al. dyed silk yarn with PEDOT:PSS and made 26 thermoelectric legs by embroidering on felted wool fabric which showed a thermoelectric voltage output of ~351 μV/K [85]. These PEDOT:PSS dyed silk yarns were stable to machine washing and up to thousand bending cycles. Recently, Jia et al. coated textile with PEDOT via VPP which combined thermoelectric generation and strain sensing application [100]. Allison et al. used vapor printing method on commercial cotton fabric to make all textile wearable band which generated thermovoltages as high as 20 mV when worn on the hand (Figure 8) [101].

Supercapacitors are alternative energy storage devices for fast charge discharge applications. Textile based supercapacitors have attracted attention due to their inherent flexibility and their potential use in wearable electronics. Nuramdhani et al. demonstrated that PEDOT:PSS sandwiched between two stainless steel conductive yarns showed capacitive behavior as an energy storage device [26]. Ma et al. reported flexible stainless steel/cotton blend yarn coated with PEDOT:PSS and PPy which can be cycles up to 5000 cycles [102]. Yuan et al. reported fiber shaped yarn supercapacitors by twisting wet spun PEDOT:PSS which showed a high areal capacitance of 119 mF/cm^2 [103]. Li et al. prepared flexible textile supercapacitors by spray coating of graphene nanosheets and PEDOT:PSS which exhibit an enhanced specific areal capacitance of 245.5 mF/cm^2 [104]. Yuksel et al. reported cotton fabric coated with MnO_2/SWNT/PEDOT:PSS ternary nanocomposite supercapacitors that gives a specific capacitance up to 246 F/g and areal capacitance of 64.5 mF/cm^2 [105].

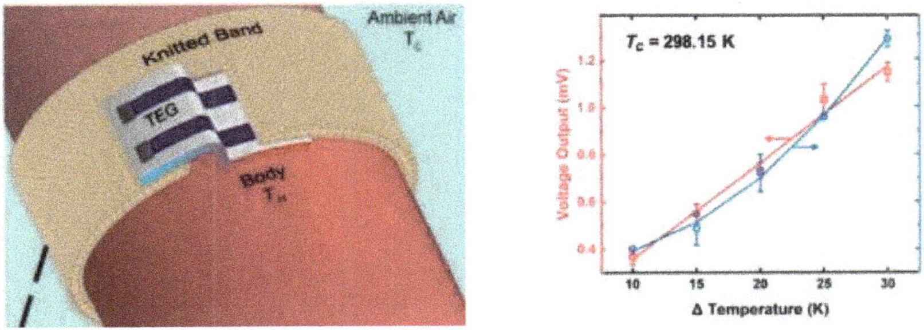

(a) (b)

Figure 8. (**a**) Design schematic of a wearable thermoelectric generator; (**b**) thermoelectric power and voltage outputs for a tobacco cotton thermopile at 25 °C. Adapted from [101].

6.3. Other Applications

There is a strong need in flexible and wearable actuators, organic light-emitting diode (OLED) and antenna. The inherent properties of PEDOT:PSS make it ideal to fabricate these devices on textiles. For instance, Li et al. developed a screen-printed textile patch antenna capable of operating at 2.4 GHz by using PEDOT:PSS as a patch and ground on polyester fabric [106]. The antenna is flexible and breathable which make it well-fit for wearable applications.

Actuation is another application area of smart textiles. Miura et al. developed a foldable PEDOT:PSS/PVA fiber by wet spinning that exhibits a repeatable contraction motion at air by applying alternating square-wave voltages between 0 and 8 V [107]. Verboven et al. reported an OLED with maintained textile properties by screen printing of silver as a bottom electrode, barium titanate as a dielectric, copper-dopped zinc-oxide as an active layer and PEDOT:PSS as a top electrode on polyester fabric that requires 3–5 V power supply [108]. The thickness of the OLED on the textile fabric was only 0.5 µm which is a good platform for wearable application; the schematic illustration and actual OLED are shown in Figure 9.

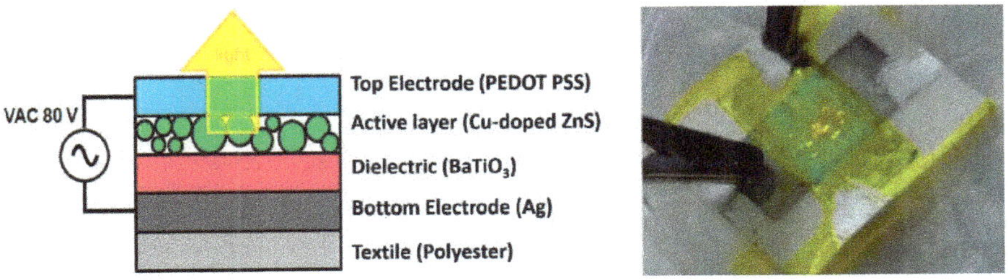

(**a**) (**b**)

Figure 9. (**a**) Build-up of the alternating current powder electroluminescence technology; (**b**) OLED stack printed on a textile. Adapted from [108].

7. Conclusions and Outlook

Current advances in textile technology, new materials, nanotechnology, and miniaturized electronics make wearable systems more feasible, but fit comfort is the ultimate key factor for wearable device user acceptance. It is convincing that this objective can only be achieved by addressing mechanical robustness and material durability in what is recognized as a harsh electronic environment: the human body and society. Thus, the use of conductive polymer composites for smart textiles could possibly be the primal solution. Composites of conductive polymers have been explored to overcome their brittleness and processability, while retaining their electrical conductivity and desirable biological properties such as cell adhesion. Enhanced mechanical properties of conductive composites usually come at the expense of desirable electrical conductivity of conductive polymers. On the other hand, fundamental understanding of the interaction between the conductive polymer filler and the non-conductive commodity polymer matrix will lead to get synergistic effect in mechanical performance and electrical properties of the composites. There is a need to achieve reasonable electrical conductivity with the lowest possible amount of conductive filler, while retaining the properties of the host polymer. The major challenges thus lie in selection of conductive filler achieving low percolation threshold and retaining biocompatibility for biomedical applications. PEDOT:PSS-based conductive polymer composites are promising for the manufacturing of smart textiles with better biocompatibility, flexibility, conductivity, printability, miscibility and weight, and as such much better suited for wearable applications compared to the common electrodes such as metallic coatings and others. As a result, tremendous PEDOT:PSS-based conductive textiles have been developed by different approaches as sensor, energy harvesting devices, antennas, OLEDs etc. However, the conductivity stability of PEDOT:PSS conductive polymer composites after being applied on textile substrates still needs an improvement. This improvement could be on the synthesis of PEDOT:PSS itself, on the combination and proportion of the polymers in the composite or by seeking new approaches of integration.

Author Contributions: G.B.T. conceived the idea; G.B.T., B.M. and D.A.M. drafted the outline; G.B.T. wrote the paper; B.M. and D.A.M. edited the paper; K.A.F. and L.V.L. supervised and administrated the project. All authors have read and agreed to the published version of the manuscript.

Funding: The research and APC were funded by the NASCERE project and the Erasmus+ project Skills4Smartex.

Acknowledgments: The authors would like to express appreciation for the support of the NASCERE and Erasmus+ project Skills4Smartex. The NASCERE project was sponsored with funds from the Ethiopian government and the Erasmus+ project Skills4Smartex has been funded with support from the European Commission. The European Commission support for the production of this publication does not constitute an endorsement of the contents which reflects the views only of the authors, and the Commission cannot be held responsible for any use which may be made of the information contained therein.

Conflicts of Interest: The authors declare no conflict of interest.

References

1. Chauraya, A.; Seager, R.; Whittow, W.; Zhang, S.; Vardaxoglou, Y. Embroidered Frequency Selective Surfaces on textiles for wearable applications. In Proceedings of the IEEE 2013 Loughborough Antennas & Propagation Conference (LAPC), Loughborough, UK, 11–12 November 2013; pp. 388–391.
2. Ding, Y.; Xu, W.; Wang, W.; Fong, H.; Zhu, Z. Scalable and Facile Preparation of Highly Stretchable Electrospun PEDOT:PSS @ PU Fibrous Nonwovens toward Wearable Conductive Textile Applications. *ACS Appl. Mater. Interfaces* **2017**. [CrossRef]
3. Seyedin, S.; Razal, J.M.; Innis, P.C.; Jeiranikhameneh, A.; Beirne, S.; Wallace, G.G. Knitted Strain Sensor Textiles of Highly Conductive All-Polymeric Fibers. *ACS Appl. Mater. Interfaces* **2015**, *7*, 21150–21158. [CrossRef] [PubMed]
4. Mikkonen, J.; Pouta, E. Flexible Wire-Component for Weaving Electronic Textiles. *Proc. Electron. Compon. Technol. Conf.* **2016**, *2016-Augus*, 1656–1663.
5. Heller, L.; Vokoun, D.; Šittner, P.; Finckh, H. 3D flexible NiTi-braided elastomer composites for smart structure applications. *Smart Mater. Struct.* **2012**, *21*, 045016. [CrossRef]

6. Shahzad, A.; Ali, Z.; Ali, U.; Khaliq, Z.; Zubair, M.; Kim, I.S.; Hussain, T.; Khan, M.Q.; Rasheed, A.; Qadir, M.B. Development and characterization of conductive ring spun hybrid yarns. *J. Text. Inst.* **2019**, *110*, 141–150. [CrossRef]
7. Okuzaki, H.; Ishihara, M. Spinning and Characterization of Conducting Microfibers. *Macromol. Rapid Commun.* **2003**, *24*, 261–264. [CrossRef]
8. Xie, J.; Pan, W.; Guo, Z.; Jiao, S.S.; Ping Yang, L. In situ polymerization of polypyrrole on cotton fabrics as flexible electrothermal materials. *J. Eng. Fibers Fabr.* **2019**, *14*, 1–8. [CrossRef]
9. Imai, T.; Shibayama, N.; Takamatsu, S.; Shiraishi, K.; Marumoto, K.; Itoh, T. High-Speed Coating Method for Photovoltaic Textiles with Closed-Type Die Coater. *Jpn. J. Appl. Phys.* **2013**, *52*, 201. [CrossRef]
10. Thibodeaux, D.P.; Baril, A. Electroless Plating of Metallic Coatings On Fabric. *J. Coat. Fibrous Mater.* **1973**, *2*, 219–230. [CrossRef]
11. Sen, P.; Xiong, Y.; Zhang, Q.; Park, S.; You, W.; Ade, H.; Kudenov, M.W.; Connor, B.T.O. Shear-Enhanced Transfer Printing of Conducting Polymer Thin Films. *ACS Appl. Mater. Interfaces* **2018**, *10*, 31560–31567. [CrossRef]
12. Tao, X. *Handbook of Smart Textiles*, 1st ed.; Tao, X., Ed.; Springer Science+ Business Media Singapore: Singapore, 2014; ISBN 9789814451680.
13. Hertleer, C.; Rogier, H.; Vallozzi, L.; Van Langenhove, L. A Textile Antenna for Off-Body Communication Integrated Into Protective Clothing for Firefighters. *IEEE Trans. Antennas Propag.* **2009**, *57*, 919–925. [CrossRef]
14. Stoppa, M.; Chiolerio, A. Wearable Electronics and Smart Textiles: A Critical Review. *Sensors* **2014**, *14*, 11957–11992. [CrossRef] [PubMed]
15. Singh, M.K. Flexible Photovoltaic Textiles for Smart Applications. In *Solar Cells - New Aspects and Solution*; Singh, M.K., Ed.; INTECH: Kanpur, India, 2011; p. 28.
16. Song, J.; Ma, G.; Qin, F.; Hu, L.; Luo, B.; Liu, T.; Yin, X.; Su, Z.; Zeng, Z.; Jiang, Y.; et al. High-Conductivity, Flexible and Transparent PEDOT:PSS Electrodes for High Performance Semi-Transparent Supercapacitors. *Polymers* **2020**, *12*, 450. [CrossRef] [PubMed]
17. Achilli, A.; Bonfiglio, A.; Pani, D. Design and characterization of screen-printed textile electrodes for ECG monitoring. *IEEE Sens. J.* **2018**, *18*, 4097–4107. [CrossRef]
18. Kuang, X.; Chen, K.; Dunn, C.K.; Wu, J.; Li, V.C.F.; Qi, H.J. 3D Printing of Highly Stretchable, Shape-Memory, and Self-Healing Elastomer toward Novel 4D Printing. *ACS Appl. Mater. Interfaces* **2018**, *10*, 7381–7388. [CrossRef]
19. Fernández-Caramés, T.; Fraga-Lamas, P. Towards The Internet-of-Smart-Clothing: A Review on IoT Wearables and Garments for Creating Intelligent Connected E-Textiles. *Electronics* **2018**, *7*, 405. [CrossRef]
20. Kirstein, T. *Multidisciplinary know-how for smart-textiles developers*, 1st ed.; Kirstein, T., Ed.; Woodhead Publishing Series in Textiles: Cambridge, UK, 2013; ISBN 978-0-85709-353-0.
21. Qu, H.; Skorobogatiy, M. Conductive polymer yarns for electronic textiles. In *Electronic Textiles*; Elsevier: Amsterdam, The Netherlands, 2015; pp. 21–53. ISBN 978-0-08-100201-8.
22. Maslik, J.; Andersson, H.; Forsberg, V.; Engholm, M.; Zhang, R.; Olin, H. PEDOT:PSS temperature sensor ink-jet printed on paper substrate. *J. Instrum.* **2018**, *13*. [CrossRef]
23. Pérez-Madrigal, M.M.; Giannotti, M.I.; Del Valle, L.J.; Franco, L.; Armelin, E.; Puiggalí, J.; Sanz, F.; Alemán, C. Thermoplastic polyurethane:polythiophene nanomembranes for biomedical and biotechnological applications. *ACS Appl. Mater. Interfaces* **2014**, *6*, 9719–9732. [CrossRef]
24. Chen, S.J.; Fumeaux, C.; Chivers, B.; Shepherd, R. A 5.8-GHz Flexible Microstrip-Fed Slot Antenna Realized in PEDOT:PSS Conductive Polymer. *2016 IEEE Int. Symp. Antennas Propag. APSURSI* **2016**, *2015*, 1317–1318.
25. Ryu, J.; Kim, J.; Oh, J.; Lim, S.; Sim, J.Y.; Jeon, J.S.; No, K.; Park, S.; Hong, S. Intrinsically stretchable multi-functional fiber with energy harvesting and strain sensing capability. *Nano Energy* **2019**, *55*, 348–353. [CrossRef]
26. Nuramdhani, I.; Jose, M.; Samyn, P.; Adriaensens, P.; Malengier, B.; Deferme, W.; De Mey, G.; Van Langenhove, L. Charge-Discharge Characteristics of Textile Energy Storage Devices Having Different PEDOT:PSS Ratios and Conductive Yarns Configuration. *Polymers* **2019**, *11*, 345. [CrossRef] [PubMed]
27. Tseghai, G.B.; Malengier, B.; Fante, K.A.; Van Langenhove, L. The Status of Textile-Based Dry EEG Electrodes. *Autex Res. J.* **2020**. Ahead of Print. [CrossRef]
28. Shi, H.; Liu, C.; Jiang, Q.; Xu, J. Effective Approaches to Improve the Electrical Conductivity of PEDOT:PSS: A Review. *Adv. Electron. Mater.* **2015**, *1*, 1500017. [CrossRef]

29. Chiang, C.K.; Fincher, C.R.; Park, Y.W.; Heeger, A.J.; Shirakawa, H.; Louis, E.J.; Gau, S.C.; MacDiarmid, A.G. Electrical Conductivity in Doped Polyacetylene. *Phys. Rev. Lett.* **1977**, *39*, 1098–1101. [CrossRef]
30. Wang, X.; Zhang, X.; Sun, L.; Lee, D.; Lee, S.; Wang, M.; Zhao, J.; Shao-Horn, Y.; Dincă, M.; Palacios, T.; et al. High electrical conductivity and carrier mobility in oCVD PEDOT thin films by engineered crystallization and acid treatment. *Sci. Adv.* **2018**, *4*, eaat5780. [CrossRef] [PubMed]
31. Shirakawa, H.; Louis, E.J.; MacDiarmid, A.G.; Chiang, C.K.; Heeger, A.J. Synthesis of electrically conducting organic polymers: Halogen derivatives of polyacetylene, (CH) x. *J. Chem. Soc. Chem. Commun.* **1977**, 578. [CrossRef]
32. Heeger, A.J. Semiconducting and Metallic Polymers: The Fourth Generation of Polymeric Materials (Nobel Lecture)*. *Angew. Chem. Int. Ed.* **2001**, 21.
33. Naarmann, H.; Theophilou, N. New process for the production of metal-like, stable polyacetylene. *Synth. Met.* **1987**, *22*, 1–8. [CrossRef]
34. Kaur, G.; Adhikari, R.; Cass, P.; Bown, M.; Gunatillake, P. Electrically conductive polymers and composites for biomedical applications. *RSC Adv.* **2015**, *5*, 37553–37567. [CrossRef]
35. Kopola, P.; Aernouts, T.; Sliz, R.; Guillerez, S.; Ylikunnari, M.; Cheyns, D.; Välimäki, M.; Tuomikoski, M.; Hast, J.; Jabbour, G.; et al. Gravure printed flexible organic photovoltaic modules. *Sol. Energy Mater. Sol. Cells* **2011**, *95*, 1344–1347. [CrossRef]
36. Mu, W.; Lin, T.; Hu, Y.; Sun, Y.; Du, Z.; Jin, J.; Zhang, D.; Cui, Z. Double layer printed high performance OLED based on PEDOT:PSS/Ir(bt) $_2$ acac:CDBP. *AIP Adv.* **2018**, *8*, 115112. [CrossRef]
37. Venkatraman, V.; Friedlein, J.T.; Giovannitti, A.; Maria, I.P.; McCulloch, I.; McLeod, R.R.; Rivnay, J. Subthreshold Operation of Organic Electrochemical Transistors for Biosignal Amplification. *Adv. Sci.* **2018**, *5*, 1800453. [CrossRef] [PubMed]
38. Taroni, P.J.; Santagiuliana, G.; Wan, K.; Calado, P.; Qiu, M.; Zhang, H.; Pugno, N.M.; Palma, M.; Stingelin-stutzman, N.; Heeney, M.; et al. Toward Stretchable Self-Powered Sensors Based on the Thermoelectric Response of PEDOT:PSS/Polyurethane Blends. *Adv Funct Mater* **2018**, *1704285*, 1–7. [CrossRef]
39. Alharbi, S.; Chaudhari, S.; Inshaar, A.; Shah, H.; Zou, C.; Harne, R.L.; Kiourti, A. E-Textile Origami Dipole Antennas With Graded Embroidery for Adaptive RF Performance. *IEEE Antennas Wirel. Propag. Lett.* **2018**, *17*, 2218–2222. [CrossRef]
40. Ramoa, S.D.A.S.; Barra, G.M.O.; Merlini, C.; Livi, S.; Soares, B.G.; Pegoretti, A. Novel electrically conductive polyurethane/montmorillonite-polypyrrole nanocomposites. *EXPRESS Polym. Lett.* **2015**, *9*, 945–958. [CrossRef]
41. Giuri, A.; Colella, S.; Listorti, A.; Rizzo, A.; Mele, C.; Esposito, C. GO/glucose/PEDOT: PSS ternary nanocomposites for flexible supercapacitors. *Compos. Part B* **2018**, *148*, 149–155. [CrossRef]
42. Qi, G.; Huang, L.; Wang, H. Highly conductive free standing polypyrrole films prepared by freezing interfacial polymerization. *Chem. Commun.* **2012**, *48*, 8246. [CrossRef]
43. Rahy, A.; Yang, D.J. Synthesis of highly conductive polyaniline nanofibers. *Mater. Lett.* **2008**, *62*, 4311–4314. [CrossRef]
44. Worfolk, B.J.; Andrews, S.C.; Park, S.; Reinspach, J.; Liu, N.; Toney, M.F.; Mannsfeld, S.C.B.; Bao, Z. Ultrahigh electrical conductivity in solution-sheared polymeric transparent films. *Proc. Natl. Acad. Sci.* **2015**, *112*, 14138–14143. [CrossRef]
45. Kirchmeyer, S.; Reuter, K. Scientific importance, properties and growing applications of poly(3,4-ethylenedioxythiophene). *J. Mater. Chem.* **2005**, *15*, 2077. [CrossRef]
46. Ouyang, J. "Secondary doping" methods to significantly enhance the conductivity of PEDOT:PSS for its application as transparent electrode of optoelectronic devices. *Displays* **2013**, *34*, 423–436. [CrossRef]
47. Palumbiny, C.M.; Liu, F.; Russell, T.P.; Hexemer, A.; Wang, C.; Müller-Buschbaum, P. The Crystallization of PEDOT:PSS Polymeric Electrodes Probed In Situ during Printing. *Adv. Mater.* **2015**, *27*, 3391–3397. [CrossRef] [PubMed]
48. Alemu, D.; Wei, H.-Y.; Ho, K.-C.; Chu, C.-W. Highly conductive PEDOT:PSS electrode by simple film treatment with methanol for ITO-free polymer solar cells. *Energy Environ. Sci.* **2012**, *5*, 9662. [CrossRef]
49. Wei, Q.; Mukaida, M.; Naitoh, Y.; Ishida, T. Morphological Change and Mobility Enhancement in PEDOT:PSS by Adding Co-solvents. *Adv. Mater.* **2013**, *25*, 2831–2836. [CrossRef] [PubMed]
50. Mengistie, D.A.; Ibrahem, M.A.; Wang, P.-C.; Chu, C.-W. Highly Conductive PEDOT:PSS Treated with Formic Acid for ITO-Free Polymer Solar Cells. *ACS Appl. Mater. Interfaces* **2014**, *6*, 2292–2299. [CrossRef]

51. Cho, B.; Park, K.S.; Baek, J.; Oh, H.S.; Koo Lee, Y.-E.; Sung, M.M. Single-Crystal Poly(3,4-ethylenedioxythiophene) Nanowires with Ultrahigh Conductivity. *Nano Lett.* **2014**, *14*, 3321–3327. [CrossRef]
52. Kroon, R.; Mengistie, D.A.; Kiefer, D.; Hynynen, J.; Ryan, J.D.; Yu, L.; Müller, C. Thermoelectric plastics: From design to synthesis, processing and structure–property relationships. *Chem. Soc. Rev.* **2016**, *45*, 6147–6164. [CrossRef]
53. Wessling, B.; Volk, H. Post-polymerization processing of conductive polymers: A way of converting conductive polymers to conductive materials? *Synth. Met.* **1986**, *15*, 183–193. [CrossRef]
54. Seyedin, M.Z.; Razal, J.M.; Innis, P.C.; Wallace, G.G. Strain-Responsive Polyurethane/PEDOT:PSS Elastomeric Composite Fibers with High Electrical Conductivity. *Adv. Funct. Mater.* **2014**, *24*, 2957–2966. [CrossRef]
55. Collier, J.H.; Camp, J.P.; Hudson, T.W.; Schmidt, C.E. Synthesis and characterization of polypyrrole-hyaluronic acid composite biomaterials for tissue engineering applications. *J. Biomed. Mater. Res.* **2000**, *50*, 574–584. [CrossRef]
56. Kim, H.S.; Hobbs, H.L.; Wang, L.; Rutten, M.J.; Wamser, C.C. Biocompatible composites of polyaniline nanofibers and collagen. *Synth. Met.* **2009**, *159*, 1313–1318. [CrossRef]
57. Lee, R.J.; Temmer, R.; Tamm, T.; Aabloo, A.; Kiefer, R. Renewable antioxidant properties of suspensible chitosan-polypyrrole composites. *React. Funct. Polym.* **2013**, *73*, 1072–1077. [CrossRef]
58. Stewart, E.M.; Fabretto, M.; Mueller, M.; Molino, P.J.; Griesser, H.J.; Short, R.D.; Wallace, G.G. Cell attachment and proliferation on high conductivity PEDOT-glycol composites produced by vapour phase polymerisation. *Biomater. Sci.* **2013**, *1*, 368–378. [CrossRef]
59. Suo, L.; Shang, X.; Tang, R.; Zhou, Y. The Preparation of Polypyrrole/Cellulose Acetate Composite Films and their Electrical Properties. In Proceedings of the ICECTT, Zhuhai, China, 31 October–1 November 2015; pp. 566–569.
60. Bajpai, A.K.; Bajpai, J.; Soni, S.N. Preparation and characterization of electrically conductive composites of poly(vinyl alcohol)-g-poly(acrylic acid) hydrogels impregnated with polyaniline (PANI). *Express Polym. Lett.* **2008**, *2*, 26–39. [CrossRef]
61. Okuzaki, H.; Takagi, S.; Hishiki, F.; Tanigawa, R. Ionic liquid/polyurethane/PEDOT:PSS composites for electro-active polymer actuators. *Sens. Actuators B Chem.* **2014**, *194*, 59–63. [CrossRef]
62. Xu, H.; Holzwarth, J.M.; Yan, Y.; Xu, P.; Zheng, H.; Yin, Y.; Li, S.; Ma, P.X. Conductive PPY/PDLLA conduit for peripheral nerve regeneration. *Biomaterials* **2014**, *35*, 225–235. [CrossRef]
63. Broda, C.R.; Lee, J.Y.; Sirivisoot, S.; Schmidt, C.E.; Harrison, B.S. A chemically polymerized electrically conducting composite of polypyrrole nanoparticles and polyurethane for tissue engineering. *J. Biomed. Mater. Res. Part A* **2011**, *98*, 509–516. [CrossRef]
64. Hilal, M.; Han, J.I. Interface engineering of G-PEDOT:PSS hole transport layer via interlayer chemical functionalization for enhanced efficiency of large-area hybrid solar cells and their charge transport investigation. *Sol. Energy* **2018**, *174*, 743–756. [CrossRef]
65. Houghton, T.; Vanjaria, J.; Yu, H. Conductive and Stretchable Silver-Polymer Blend for Electronic Applications. *2016 IEEE 66th Electron. Compon. Technol. Conf. ECTC* **2016**, 812–816.
66. Zajdel, T.J.; Baruch, M.; Méhes, G.; Stavrinidou, E.; Berggren, M.; Maharbiz, M.M.; Simon, D.T.; Ajo-Franklin, C.M. PEDOT:PSS-based Multilayer Bacterial-Composite Films for Bioelectronics. *Sci. Rep.* **2018**, *8*, 15293. [CrossRef]
67. Hebbar, V.; Bhajantri, R.F.; Ravikumar, H.B.; Ningaraju, S. Role of free volumes in conducting properties of GO and rGO filled PVA- PEDOT: PSS composite free standing films: A positron annihilation lifetime study. *J. Phys. Chem. Solids* **2019**, *126*, 242–256. [CrossRef]
68. Zahid, M.; Papadopoulou, E.L.; Athanassiou, A.; Bayer, I.S. Strain-responsive mercerized conductive cotton fabrics based on PEDOT:PSS/graphene. *Mater. Des.* **2017**, *135*, 213–222. [CrossRef]
69. Chen, M.; Duan, S.; Zhang, L.; Wang, Z.; Li, C. conductive polymer composites based on graphene networks grown by chemical vapour deposition and PEDOT:PSS coating. *Chem. Commun.* **2015**, 3169–3172. [CrossRef] [PubMed]
70. Meng, Q.; Cai, K.; Du, Y.; Chen, L. Preparation and thermoelectric properties of SWCNT/PEDOT: PSS coated tellurium nanorod composite films. *J. Alloys Compd.* **2019**, *778*, 163–169. [CrossRef]
71. Jalili, R.; Razal, J.M.; Innis, P.C.; Wallace, G.G. One-Step Wet-Spinning Process of Poly (3,4-ethylenedioxythiophene): Poly (styrenesulfonate) Fibers and the Origin of Higher Electrical Conductivity. *Adv. Funct. Mater.* **2011**, 3363–3370. [CrossRef]

72. Okuzaki, H.; Harashina, Y.; Yan, H. Highly conductive PEDOT/PSS microfibers fabricated by wet-spinning and dip-treatment in ethylene glycol. *Eur. Polym. J.* **2009**, *45*, 256–261. [CrossRef]
73. Zhou, J.; Li, E.Q.; Li, R.; Xu, X.; Ventura, I.A.; Moussawi, A.; Anjum, D.H.; Hedhili, M.N.; Smilgies, D.M.; Lubineau, G.; et al. Semi-metallic, strong and stretchable wet-spun conjugated polymer microfibers. *J. Mater. Chem. C* **2015**, *3*, 2528–2538. [CrossRef]
74. Zhang, J.; Seyedin, S.; Qin, S.; Lynch, P.A.; Wang, Z.; Wang, X.; Razal, J.M.; Yang, W. Fast and scalable wet-spinning of highly conductive PEDOT:PSS fibers enables versatile applications. *J. Mater. Chem. A* **2019**, 6401–6410. [CrossRef]
75. Liu, Y.; Li, X.; Lu, J.C. Electrically Conductive Poly (3,4-ethylenedioxythiophene)– Polystyrene Sulfonic Acid/Polyacrylonitrile Composite Fibers Prepared by Wet Spinning. *J. Appl. Polym. Sci.* **2013**, 370–374. [CrossRef]
76. Jin, S.; Sun, T.; Fan, Y.; Wang, L.; Zhu, M.; Yang, J.; Jiang, W. Synthesis of freestanding PEDOT:PSS/PVA@Ag NPs nanofiber film for high-performance flexible thermoelectric generator. *Polymer* **2019**, *167*, 102–108. [CrossRef]
77. Zhang, Q.; Wang, X.; Fu, J.; Liu, R.; He, H.; Ma, J.; Yu, M.; Ramakrishna, S.; Long, Y. Electrospinning of Ultrafine Conducting Polymer Composite Nanofibers with Diameter Less than 70 nm as High Sensitive Gas Sensor. *Materials* **2018**, *11*, 1744. [CrossRef] [PubMed]
78. Allison, L.; Hoxie, S.; Andrew, T.L. Towards Seamlessly-Integrated Textile Electronics: Methods to Coat Fabrics and Fibers with Conducting Polymers for Electronic Applications. *Chem. Commun.* **2017**. [CrossRef]
79. Hong, K.H.; Oh, K.W.; Kang, T.J. Preparation and properties of electrically conducting textiles by in situ polymerization of poly(3,4-ethylenedioxythiophene). *J. Appl. Polym. Sci.* **2005**, *97*, 1326–1332. [CrossRef]
80. Bashir, T.; Skrifvars, M.; Persson, N.-K. High-strength electrically conductive fibers: Functionalization of polyamide, aramid, and polyester fibers with PEDOT polymer. *Polym. Adv. Technol.* **2018**, *29*, 310–318. [CrossRef]
81. Bashir, T.; Ali, M.; Cho, S.-W.; Persson, N.-K.; Skrifvars, M. OCVD polymerization of PEDOT: Effect of pre-treatment steps on PEDOT-coated conductive fibers and a morphological study of PEDOT distribution on textile yarns: OPTIMIZATION OF CVD PROCESS AND SURFACE MORPHOLOGY. *Polym. Adv. Technol.* **2013**, *24*, 210–219. [CrossRef]
82. Trindade, I.G.; Martins, F.; Baptista, P. High electrical conductance poly(3,4-ethylenedioxythiophene) coatings on textile for electrocardiogram monitoring. *Synth. Met.* **2015**, *210*, 179–185. [CrossRef]
83. Zhang, L.; Fairbanks, M.; Andrew, T.L. Rugged Textile Electrodes for Wearable Devices Obtained by Vapor Coating Off-the-Shelf, Plain-Woven Fabrics. *Adv. Funct. Mater.* **2017**, *27*, 1700415. [CrossRef]
84. Ding, Y.; Invernale, M.A.; Sotzing, G.A. Conductivity trends of pedot-pss impregnated fabric and the effect of conductivity on electrochromic textile. *ACS Appl. Mater. Interfaces* **2010**, *2*, 1588–1593. [CrossRef]
85. Ryan, J.D.; Mengistie, D.A.; Gabrielsson, R.; Lund, A.; Müller, C. Machine-Washable PEDOT:PSS Dyed Silk Yarns for Electronic Textiles. *ACS Appl. Mater. Interfaces* **2017**, *9*, 9045–9050. [CrossRef]
86. Lund, A.; Darabi, S.; Hultmark, S.; Ryan, J.D.; Andersson, B.; Ström, A.; Müller, C. Roll-to-Roll Dyed Conducting Silk Yarns: A Versatile Material for E-Textile Devices. *Adv. Mater. Technol.* **2018**, *3*, 1800251. [CrossRef]
87. Tadesse, M.G.; Mengistie, D.A.; Chen, Y.; Wang, L.; Loghin, C.; Nierstrasz, V. Electrically conductive highly elastic polyamide/lycra fabric treated with PEDOT:PSS and polyurethane. *J. Mater. Sci.* **2019**, *54*, 9591–9602. [CrossRef]
88. Guo, Y.; Otley, M.T.; Li, M.; Zhang, X.; Sinha, S.K.; Treich, G.M.; Sotzing, G.A. PEDOT:PSS "wires" Printed on Textile for Wearable Electronics. *ACS Appl. Mater. Interfaces* **2016**, *8*, 26998–27005. [CrossRef] [PubMed]
89. Sinha, S.K.; Noh, Y.; Reljin, N.; Treich, G.M.; Hajeb-mohammadalipour, S.; Guo, Y.; Chon, K.H.; Sotzing, G.A. Screen-Printed PEDOT:PSS Electrodes on Commercial Finished Textiles for Electrocardiography. *ACS Appl. Mater. Interfaces* **2017**, *9*, 37524–37528. [CrossRef] [PubMed]
90. Zhao, X.; Ding, J.; Bai, W.; Wang, Y.; Yan, Y.; Cheng, Y. PEDOT: PSS/AuNPs/CA modi fi ed screen-printed carbon based disposable electrochemical sensor for sensitive and selective determination of carmine. *J. Electroanal. Chem.* **2018**, *824*, 14–21. [CrossRef]
91. Tseghai, G.B.; Malengier, B.; Fante, K.A.; Nigusse, A.B.; Van Langenhove, L. Development of a Flex and Stretchy Conductive Cotton Fabric Via Flat Screen Printing of PEDOT:PSS/PDMS Conductive Polymer Composite. *Sensors* **2020**, *20*, 1742. [CrossRef]

92. Kang, T.-K. Piezoresistive Characteristics of Nylon Thread Resistive Memories for Wearable Strain Sensors. *Coatings* **2019**, *9*, 623. [CrossRef]
93. Pani, D.; Dess, A.; Saenz-cogollo, J.F.; Barabino, G.; Fraboni, B.; Bonfiglio, A. Fully Textile, PEDOT:PSS Based Electrodes for Wearable ECG Monitoring Systems. *IEEE Trans. Biomed. Eng.* **2016**, *63*, 540–549. [CrossRef]
94. Ankhili, A.; Tao, X.; Koncar, V.; Coulon, D.; Tarlet, J. Ambulatory Evaluation of ECG Signals Obtained Using Washable Textile-Based Electrodes Made with. *Sensors* **2019**, *19*, 416. [CrossRef]
95. Ankhili, A.; Tao, X.; Cochrane, C.; Koncar, V.; Coulon, D.; Tarlet, J.-M. Comparative Study on Conductive Knitted Fabric Electrodes for Long-Term Electrocardiography Monitoring: Silver-Plated and PEDOT:PSS Coated Fabrics. *Sensors* **2018**, *18*, 3890. [CrossRef]
96. Niijima, A.; Isezaki, T.; Aoki, R.; Watanabe, T. hitoeCap: Wearable EMG Sensor for Monitoring Masticatory Muscles with PEDOT-PSS Textile Electrodes. In Proceedings of the ISWC '17, Maui, HI, USA, 11–15 Sepember 2017; pp. 215–220.
97. Abbasi, M.A.B.; Vryonides, P.; Nikolaou, S. Humidity Sensor Devices using PEDOT: PSS. In Proceedings of the 2015 IEEE International Symposium on Antennas and Propagation & USNC/URSI National Radio Science Meeting, Vancouver, BC, Canada, 19–24 July 2015; pp. 1366–1367.
98. Smith, R.E.; Totti, S.; Velliou, E.; Campagnolo, P.; Hingley-Wilson, S.M.; Ward, N.I.; Varcoe, J.R.; Crean, C. Development of a novel highly conductive and flexible cotton yarn for wearable pH sensor technology. *Sens. Actuators B Chem.* **2019**, *287*, 338–345. [CrossRef]
99. Du, Y.; Cai, K.; Chen, S.; Wang, H.; Shen, S.Z.; Donelson, R.; Lin, T. Thermoelectric Fabrics: Toward Power Generating Clothing. *Sci. Rep.* **2015**, *5*, 6411. [CrossRef] [PubMed]
100. Jia, X.; Tennant, A.; Langley, R.J.; Hurley, W.; Dias, T. A knitted textile waveguide. In Proceedings of the IEEE 2014 Loughborough Antennas and Propagation Conference (LAPC), Loughborough, UK, 10–11 November 2014; pp. 679–682.
101. Allison, L.K.; Andrew, T.L. A Wearable All-Fabric Thermoelectric Generator. *Adv. Mater. Technol.* **2019**, *4*, 1800615. [CrossRef]
102. Ma, Y.; Wang, Q.; Liang, X.; Zhang, D.; Miao, M. Wearable supercapacitors based on conductive cotton yarns. *J. Mater. Sci.* **2018**, *53*, 14586–14597. [CrossRef]
103. Yuan, D.; Li, B.; Cheng, J.; Guan, Q.; Wang, Z.; Ni, W.; Li, C.; Liu, H.; Wang, B. Twisted yarns for fiber-shaped supercapacitors based on wetspun PEDOT:PSS fibers from aqueous coagulation. *J. Mater. Chem. A* **2016**, *4*, 11616–11624. [CrossRef]
104. Li, Z.; Ma, Y.; Wang, L.; Du, X.; Zhu, S.; Zhang, X.; Qu, L.; Tian, M. Multidimensional Hierarchical Fabric-Based Supercapacitor with Bionic Fiber Microarrays for Smart Wearable Electronic Textiles. *ACS Appl. Mater. Interfaces* **2019**, *11*, 46278–46285. [CrossRef]
105. Yuksel, R.; Unalan, H.E. Textile supercapacitors-based on MnO_2/SWNT/conducting polymer ternary composites: Textile-based supercapacitors. *Int. J. Energy Res.* **2015**, *39*, 2042–2052. [CrossRef]
106. Li, Z.; Sinha, S.K.; Treich, G.M.; Wang, Y.; Yang, Q.; Deshmukh, A.A.; Sotzing, G.A.; Cao, Y. All-organic flexible fabric antenna for wearable electronics. *J. Mater. Chem. C* **2020**. [CrossRef]
107. Miura, H.; Fukuyama, Y.; Sunda, T.; Lin, B.; Zhou, J.; Takizawa, J.; Ohmori, A.; Kimura, M. Foldable Textile Electronic Devices Using All-Organic Conductive Fibers: Foldable Textile Electronic Devices Using ... . *Adv. Eng. Mater.* **2014**, *16*, 550–555. [CrossRef]
108. Verboven, I.; Stryckers, J.; Mecnika, V.; Vandevenne, G.; Jose, M.; Deferme, W. Printing Smart Designs of Light Emitting Devices with Maintained Textile Properties. *Materials* **2018**, *11*, 290. [CrossRef]

MDPI AG
Grosspeteranlage 5
4052 Basel
Switzerland
Tel.: +41 61 683 77 34

Sensors Editorial Office
E-mail: sensors@mdpi.com
www.mdpi.com/journal/sensors

www.ingramcontent.com/pod-product-compliance
Lightning Source LLC
Chambersburg PA
CBHW062056220926
51850CB00010B/2791
* 9 7 8 3 7 2 5 8 7 6 9 1 4 *